WARPLANE *Plus* NO. 01

A-17

THE COMPLETE HISTORY OF THE NORTHROP ATTACK PLANES AND ITS EXPORT DERIVATIVES

BY SANTIAGO RIVAS, AMARU TINCOPA, NICO BRAAS AND EDWIN HOOGSCHAGEN

Violaero

JACK NORTHROP'S EARLY AIRCRAFT

ABOUT ALPHA, BETA, GAMMA AND DELTA

By Nico Braas

Aircraft constructor and manufacturer Jack Northrop is well known for his flying wing designs. Lesser known is the story of his earlier designs.

Born as John Knudsen Northrop on 10 November 1895 he started his aviation career in 1916 at Loughead Aircraft Manufacturing. When this company folded, he joined the Douglas Aircraft Company as a project engineer where he worked together with Douglas designer Edward Heinemann. Northrop had always been interested in building light weight monocoque fuselages and in his free hours he designed his very first aircraft. He left Douglas and rejoined the newly formed Lockheed Aircraft Company. Here, his design would materialise as the Lockheed Vega. After a short spell at Lockheed, he left again. He started his own company in 1929; the Avion Corporation. He sold this company two years later and in 1932 he established a company under his own name: the Northrop Corporation. It was a joint-venture with Donald Douglas with its main production facilities at El Segundo, California. Here, he started with the design and construction of a small range of modern looking and for that time very revolutionary monoplanes for civil and military use as the Alpha, Beta, Gamma and Delta.

NORTHROP ALPHA

The first type built at the El Segundo plant was a very modern all-metal low-wing monoplane known as the 'Alpha'.

It was powered by a single 420 hp Pratt & Whitney Wasp air-cooled radial engine. It had, directly after the engine firewall, a fully enclosed cabin with capacity for six passengers.

Violaero

© Copyright 2019 Lanasta, Odoorn

www.lanasta.com

Editor
Edwin Hoogschagen

Author
Santiago Rivas
Amaru Tincopa
Nico Braas
Edwin Hoogschagen

Corrections
Mats Averkvist,
Edwin Hoogschagen,
Rob Mulder

Graphic design
Jantinus Mulder

Publisher
Lanasta

First print, October 2019
ISBN 978-90-8616-271-0
NUR 465

Contact Warplane Plus:
Oude Kampenweg 29,
7873 AG Odoorn (NL)
Tel. 0031 (0)591 618 747
info@lanasta.eu

Alpha 2 NC999Y was the third machine delivered to TWA. (coll. N. Braas)

Based on the latest wind tunnel findings and latest 'state of art' of aeronautical engineering it featured an aluminium monocoque fuselage and an all-metal, unbraced 'stressed-skin' wing with a de-icing device on the wing leading edge. The main wheels of the undercarriage were uncovered, but later types had slim trouser-type wheelspats. In spite of its for that time very modern construction, the cockpit was still open with only a small front screen.

The first flight of the prototype with registration *X-2W* took place in early 1930 with test pilot Eddy Allen at the controls. Unfortunately, it already crashed on its second flight when it lost an aileron. The first Alpha 2 carried the registration *X127W*. In spite of this five were ordered and put into service as Alpha *2* at Transcontinental & Western Air or TWA. Later, this company became well-known as Trans World Airlines! In total, TWA used thirteen Alpha's on its regular flying routes from San Francisco to New York until phased out in 1935. With a total of 13 intermediate stops it could make this flight in just 23 hours. Some Alpha 2s were converted into

The second built Alpha 2. (coll. N. Braas)

a two-seat version with some cargo capacity for freight and airmail as Alpha 3. A further conversion for freight only was the Alpha 4 with a slightly larger wing and streamlined main wheel covers.

One Alpha, *NC999Y* was used by the NACA flight test centre at Langley.

One Alpha carried the registration *NR11Y* as a racing plane and it flew not only with a wheel based undercarriage but also on floats. It seems this registration never was official (in fact it was *NC11Y*) and it never entered any racing contest. *NR11Y* still exists at the National Air & Space Museum.

The US Army Air Corp Y1C-19 with serial 31-518.

(coll. N. Braas)

The Army Air Force used one Alpha 1 as the YC-19 and two as Y1C-19 for military VIP transport, with the passenger capacity reduced to four. Their serial numbers were *31-316, -317* and *-318.* The first YC-19 was delivered in May 1931, and used as a fast staff transport primarily for Army Generals and Washington, D.C. and VIPs (i.e. politicians). From the summer of 1931 until the spring of 1933, the C-19s, based at Wright and Bolling Fields, shuttled VIPs to various locations around the country. Y1C-19 *31-318* was destroyed on March 20, 1933 in a fatal crash, the remaining planes were removed from VIP service and sent to regular Air Corps units for use as squadron 'hacks', light cargo and military passenger transports. The remaining two C-19s were kept in service until mid-1939 when both planes were sent to aircraft mechanic schools and used as ground trainers. Total number of Alpha's built was 17.

The second, still unmarked, Northrop Beta.

(coll. N. Braas)

NORTHROP BETA

Roughly based on the earlier Alpha, Northrop designed and built a smaller all-metal aircraft as the 'Beta.' Responsible for its design was Don Berlin (who later became chief designer at Curtiss). The Beta c/n 1 was a low wing monoplane of all-metal construction with two separate

cockpits and well-streamlined trouser fairings around its main wheels. For its time it was a very advanced construction fitted with a 160 hp Menasco Buccaneer inline air-cooled engine. The first flight was made on 3 March 1931. Initially it carried the civil registration *X963Y* (later *NX963Y*) but when it received its full Certificate of Airworthiness this was changed into

NC963Y. It did not last very long; on 12 August 1931 it crashed at Los Angeles.

At Northrop a second Beta c/n 2 was built that was quite different from the first one. The second cockpit was deleted and as a single-seater it was fitted with a much more powerful engine: a 300 hp Pratt & Whitney Wasp Jr. radial air-cooled engine. Carrying the registration *X12214* (later changed into *N12214*) it could reach a speed in excess of 322 km/h (200 mph).

Stearman was at that time Northrop's sister company based at Wichita and here it was used as a demonstrator for possible clients. Because of the economical crisis at that time not a single one was ordered and *N12214* was finally sold to a wealthy private pilot. However, during its delivery to New York it made an intermediate landing at Wright Field where it was thoroughly tested by the Army Air Corps since at that time the A.A.C. still used biplanes! It was rarely flown by its new owner and in 1932

the plane changed hands again to a new owner who based it at Roosevelt Field. It was damaged later at a nearby airfield. It was repaired and used by Stearman at Wichita as an experimental test platform for various flap designs until it crashed on 5 May 1934 due to a structural failure of a wing.

The Beta, designed as two seat sports plane.
(coll. N. Braas)

The Beta, with X-registry. The Stearman company would dub it Model 3D.
(coll. N. Braas)

Northrop Gamma

Early 1932 Northrop started with their next type in the Greek alphabet: the Gamma. It was a high-performance aircraft built according to the latest techniques i.e. all metal, unbraced sleek monoplane with a monocoque fuselage and stressed-skin wings. It was intended as a fast mail carrier, but the performances were so high that it was also used for racing and long-distance flights. It featured the same type of fixed undercarriage as the Alpha with well-streamlined trouser-type fairings. It also featured a fully enclosed two-seat cockpit.

Power was provided by a 785 hp Wright Whirlwind radial engine. What Jack Northrop also had in mind was that it could easily be developed further for military use. Built at Northrop's new Mines Field production facility, the first Gamma was designated as the Gamma 2A with registration *X12265*. Later this was changed into *NR12265*. It was built as a single seater, owned by Texas Oil Company (Texaco) and flown by the famous pilot Frank Hawks in some record breaking flights. Named *Sky Chief* it made on 2 June 1933 a non-stop flight from Los Angeles to

New York in 13 hours 27 minutes with an average speed of 291 km/h. In 1934, it was entered in the Bendix Trophy Race but it crashed after an in-flight fire had broken out.

The second Gamma (the 2B) was also purely intended for record breaking flights. Carrying the registration *NR122269* it was built as a two-seater for Lincoln Ellsworth. It was named *Polar Star* and fitted with a 500 hp Pratt & Whitney Wasp. It could be fitted with the streamlined wheels undercarriage and with twin floats made by Edo.

TWA purchased three Gamma's, known as 2D.
(coll. N. Braas)

If needed the wheels could be replaced by skis. Purpose of this special built version was to make flights across the Antarctic continent. After two failed expeditions in 1934 and early 1935, a third attempt was made in November 1935. It ended after some thousands kilometres on Antarctica with an emergency landing without fuel. Ellsworth and his plane were picked up by a British ship. *Polar Star* eventually ended in the collection of the National Air and Space Museum.

A further three Gamma's (c/n nos. 8-10) were ordered by TWA as Gamma 2D. They carried the civil registrations *X/NR13757,*

Jacqueline Cochrane for her entry in the McRoberston Race from the U.K. to Australia in October 1934. The plane was already damaged during its delivery flight and Jacqueline Cochrane eventually entered the McRobertson Race with a two-seat Gee Bee racer (without success since she had to give up at Budapest!). *NC13671* was later rebuilt and fitted with a Pratt & Whitney radial engine for the Bendix Trophy Race in 1935. The plane had to give up this prestigious race because of deteriorating weather conditions. Re-engined with a 1000 hp Wright SR-1820-G2 radial engine it was leased to Howard Hughes. He set a new transcontinental non-stop record with this aircraft by covering Burbank to

named Delta, a single 710 hp Wright SR-1820 engine. It had accommodation for six passengers. First flight with c/n 3 *X/NC12292* was made in May 1933. It was lost in a crash in October 1933 after it had been damaged at an earlier landing incident. The next Delta was powered by a 660 hp Pratt and Whitney Hornet engine. Unfortunately for Northrop a restriction was put on single engine aircraft used for passenger operations at night and over difficult terrain. For this reason only a small number was built that saw mostly use as executive transport in (wealthy) private hands. Most Delta's were powered again by the more powerful SR-1820 engine, but a few received also the P&W Hornet.

Jacqueline Cochran's streamlined Gamma 2G.
(coll. N. Braas)

X/NR13758 and *NC13759.* They were purchased and delivered in 1934 as fast mail planes with capacity of 633 kg in two fuselage cargo holds. The first TWA Gamma 2D was used for record breaking transcontinental flights as *NC13757.* With this aircraft TWA pilot Jack Fry established a transcontinental one-stop record on 13-14 May 1934, by covering Los Angeles-Kansas City- Newark in 11 hr 31 min with a 152 kg payload of special mail. Another TWA Gamma 2D, *X13758,* was later used for high-altitude flight testing as *NR13758.*
Gamma c/n 10 was purchased by Texaco for promotional flights in 1935. It ended its day as a bomber during the Spanish Civil War.

Gamma 2G c/n 11 *X13671/NC13761* was fitted with a 700 hp Curtiss Conqueror liquid-cooled engine ordered by

Newark in 9 hr 26 min at an average speed of 417 km/h.
Gamma 2H c/n 12 *X2111/NR2111* was used for racing purposes by respectively Marion Guggenheim, Russel Thaw and Bernard McFadden.

NORTHROP DELTA

In August 1932, when the first Gamma was completed, Northrop developed in parallel a larger type which better suited the needs of the airline companies. Northrop selected for his new all-metal airliner,

One Delta (c/n 29)was quite different in appearance fitted with a rearward placed Gamma cockpit. It flew at the Swedish company AB Aerotransport with registration *SE-ADW* carrying the name Smâlland. AB Aerotransport also operated a second standard Delta (c/n 7) with registration *SE-ADI* named Halland.
Two Delta's were used in the Spanish Civil War by the Nationalists carrying the registrations *43-4* and *43-5.* as light transport planes.
One Delta, c/n 74, was used by the U.S. Coast Guard as RT-1 with registration *V150.*

Technical data				
	Alpha 2	Beta 2	Gamma 2D	Delta 1B
Engine	P&W Wasp	P&W Wasp Jr	Wright SR-1820	P & W Hornet
Power	420 hp	300 hp	710 hp	660 hp
Wingspan	12.8 m	9,75 m	14.57 m	14.55 m
Length	8.7 m	6,6 m	9.50 m	10.44 m
Height	2.7 m		2.74 m	2.95 m
Empty weight	1177 kg		1868 kg	1860 kg
All-up weight	2045 kg		3334 kg	3175 kg
Max. speed	285 km/h	341 km/h	359 km/h	338 km/h
Range	2650 km		3170 km	2495 km
Service ceiling	5885 m		7130 m	6095 m
Accommodation:	1 + 6	pilot only	1 or 2	1 + 6

A MILITARY GAMMA FAMILY

Construction of the third Gamma (c/n 5) started as private venture and was completed in May 1933 as Gamma 2C. It was sent to the Army Air Corps for evaluation at Wright Field. The plane was armed with four fixed machineguns in the wings plus a fifth trainable example in the observers' cockpit. An 1100 lbs bombload could be carried externally. Power was provided by a 735 hp Wright SR-1820-F2 radial engine. It carried the civil registration *X12291*, and flew for the first time during spring of 1933. During the following period, it was thoroughly tested by the Army Air Corps before returning to Northrop for modification work in February 1934. The tail planes were replaced by slightly larger, more triangular shaped examples. The plane re-

The last Delta built, c/n 185 was shipped to Canada as a pattern aircraft for licence production.

Total number of Delta's built was 13 with another 19 built under a licence by Canadian Vickers for use by the Royal Canadian Air Force. Canadian Delta's were sometimes fitted with snow skids and at least one was fitted with floats.

Right: *An impressive front view of the YA-13, showing its thick wing and massive engine which was much wider than the fuselage.* (coll. N. Braas)

The XA-16 with its smaller diameter 14 cylinder Pratt & Whitney R-1830-7 engine. (coll. N. Braas)

ceived the new military designation YA-13 and was purchased on 28 June of the same year, at a cost of $ 80,950, receiving military serial *34-027*.

The XFT-2 was Northrops second attempt to develop a naval fighter. After its crash, the project was shelved. (coll. N. Braas)

It was found that the large diameter engine interfered with forward vision and the plane, after logging 22 flying hours, was again returned to the Northrop factory in January 1935 for modifications. The engine was replaced by a Pratt & Whitney R-1830-7 Twin Wasp of 950 hp of smaller diameter. The machine re-emerged with the new Army Air Corps type designation XA-16 and was sent back to Wright Field for further evaluation on 18 April 1935. After initial testing, the new engine proved to be too powerful, and further refinements were needed. the XA-16 program was ended after its testing period was completed, having accumulated 732 flight hours. It ended up as instructional airframe at an aircraft mechanics' school at Roosevelt Field, London Island.

Navy projects

Parallel to the Army Air Corps projects, a line of navalized planes took shape. C/N 6 was a single seat shipboard fighter designated as XFT-1. In fact, it was only roughly based on the Gamma. It was powered by a Wright R-1510-26 air-cooled radial engine of 625 hp, received U.S. Navy BuNo. 9400 and was delivered to Anacostia Naval Air Station on 14 February 1934. It was rejected by the Navy, even after it had

been fitted with a more powerful 700 hp R-1535 engine and redesignated as XFT-2. Flight characteristics were still unacceptable and the XFT-2 was returned to the factory. During its return flight it crashed over the Allegheny Mountains, and further developments were terminated.

Export to China

Construction nos. 14-27, 30-37 and 45, 46 were Gamma's 2E ordered by the Chinese government. The series production was a welcome source of income for the compa-

ny. They were fitted with military equipment including a flexible machine gun in the rear cockpit and four fixed machine guns in the wings. They could be used for armed reconnaissance and as light bomber. They were powered by a 710 hp Wright SR-1820-F3 driving two-blade propellers. Total supply was 24 aircraft. All were delivered between February and December 1934. Another 25 (c/n 48-72) were delivered as components for local assembly in China by the Central Aircraft Manufacturing Company (CAMCO).

An unmarked Gamma 2E, the prototype of a series intended for China. (coll. N. Braas)

During their operational use, the aircraft still left after numerous training accidents were destroyed during the early phase of the military conflicts with Japan.

A single Gamma 2E was purchased by the British Air Ministry for evaluation at Martlesham Heath test centre. C/n 13 received R.A.F. colors and serial *K5053*. Conclusion here was that 'performance was not outstanding' and after some use as a general purpose aircraft it quietly disappeared and its final fate is unknown.

to replace the dive brakes with new units which were perforated. This indeed solved the buffeting problems. The prototype was accepted on 12 December and an order for 54 production BT-1s powered by a 825 hp P & W R-1535-94 radial engine was placed in September 1936.

Gamma c/n 47 was designated as Gamma 2ED-C. It was powered by a 735 hp Wright SR-1820-F53 engine. It was completed in 1934 as a demonstration aircraft with civil registration *X13670*. During 1935 it made an extensive demonstration tour in Central and South America, flown by Frank

Hawks and G.H. Irving. Total flying time was 101.23 hours at an average speed of some 320 km/h. It was flown back to Los Angeles in May 1935 and was finally sold to the Soviet Union in November 1935.

NAVY DIVE BOMBER

Another offspring of the Gamma line was a light naval bomber and attack plane. It was designed in response to a specification drawn up in June 1934. The Northrop project received preference by the US Navy and Northrop was awarded a contract for a single prototype under designation XBT-1 in November 1934. Technically, it was a scaled down, refined and strengthened model 2C. It featured main landing gear legs that retracted backwards in two streamlined underwing fairings. The wheels were only partly retracted for making emergency landings with minimal damage. Further, the BT-1 was fitted with an arrester hook for operations from aircraft carriers. The prototype (C/n 43, BuNo. 9745) made its first flight on 19 August 1935. After initial flight testing severe buffeting was experienced during test dives. To cure this, NACA recommended

The single Gamma 2E delivered to Great Britain. (coll. N. Braas)

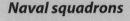

Naval squadrons

The BT-1s were assigned to naval squadrons VB-5 (operational on USS Yorktown) and VB-6 (on USS Enterprise) and were used until 1941. A further development was the BT-2 with a fully revised fuselage and a 1000 hp Wright R-1820-G133 engine and an undercarriage that was fully retractable in the wings. First flight of the XBT-2, still fitted with a 825 hp P & W R-1535 engine, took place in April 1938. The U.S. Navy was very impressed with the performances of the BT-2 and soon an order followed. The first five were still 'Northrops', but with the El Segundo plant fully taken over in 1939 by Douglas they were from that moment Douglas planes. The type designation of the BT-2 was soon changed into SBD-1 it received the name 'Dauntless'. SBD-1 and 2s were the most important dive bombers of the U.S. Navy when the war with Japan started and the type played a major role in the Second World War!

A BT-1 at El Segundo. The revised XBT-2 variant was to be the forerunner of the famous Dauntless divebomber. (coll. N. Braas)

An Army Air Corps order

During the trial and error experiences with the Gamma 2C, dubbed YA-13 and XA-16, Northrop had already a much improved parallel type under development. This was the Gamma 2F, with a slimmer fuselage and 750 HP Pratt & Whitney R-1535-11 fourteen-cylinder radial engine. The cockpit was not only lowered, it carried a rounded, much more streamlined windscreen and hood over the pilot's cockpit. The entire cockpit area was also lowered and placed further behind. The tail surfaces were again redesigned. While the 2C featured a horizontal tail plane located at the centre line of the fuselage, the new 2F featured tail planes which were relocated to a new higher location. This relocation was recommended by Material Division of the Air Corps, in order to improve the spin characteristics of the new design. As a novelty, the new 2F was fitted with a retractable main landing gear. The gear legs retracted rearward into large fairings. Evaluation started on 6 October 1934. The Army Air Corps reported that the machine was satisfactory from an engineering standpoint, but had to be significantly altered if it was to be satisfactory for the purpose of attack aviation. Both projects looked very promising. Just a short while later, Northrop was awarded a contract to build one sample of the XBT-1 dive bomber.

Both photos: *The model 2F was a much more aesthetically pleasing machine compared to the earlier Northrop attack projects. Here, the machine is still fitted with the much criticised retractable landing gear.* (coll. G. Balzer)

able to properly operate the rear gun. This area, including the rear gunner's canopy, had to be redesigned. A turnover structure should be added between front and rear cockpits to protect the crew if the plane should turn over in an accident. Besides the arrangement of the cockpit areas, the bulky, partially retractable landing gear attracted a lot of attention. The reviewing committee was unanimous in its verdict, that this type of landing gear was unacceptable. It was considered to be dangerous in the event of a forced landing. Either a fully retractable or fixed landing gear was preferred. Given the long list of recommendations, the committee was concerned about

Two lengthy reports on the Gamma 2F were compiled, dated 11 and 27 October. The first was written after a group of pilots was given the opportunity to fly the plane. The second was written by lt. col. Horace M. Hickam, president of the committee studying possible aircraft purchases. Among the comments was the enclosed pilots' cockpit - an open cockpit was still preferred. Cockpit layout itself was also criticized; the basic controls, including flap, stick, propeller pitch and trim control were not ergonomically produced, brake pedal activation should be changed to the standard Army Air Corps system, and lastly, redesign of all instrument panels was strongly advised. The rear cockpit was also reviewed. Main concern was that this area was rather confined. A crewman wearing thick winter clothing would not be

Another set of images reveals model 2F with fixed main landing gear struts.
(coll. G. Balzer)

A walk around the prototype reveal a lot of details which were modified in the production aircraft. The triangular shaped tail fin still needs to be enlarged, and the whole area between engine cowling and cockpit hood will be further refined, including the air scoop at the port side of the fuselage and oil cooler intake at the bottom of the fuselage. (Coll. G. Balzer)

35-54, just after delivery to 3rd AG. This took place on 2 February 1936. At left the nose of a Curtiss A-12. (coll. G. Balzer)

two points. First, it was to be expected that Northrop would not be able to offer the design at the same cost after all modification work would be completed. Secondly, performance of the airplane could be negatively affected after changing the landing gear and deleting the closed cabins.

The Army Air Corps was confident, however, that the new Northrop type was a promising aircraft, and on 24 December 1934 it was announced to Northrop that the Army Air Corps had the intention to purchase 110 series aircraft after modification work would be completed. The small Northrop company had to bring the A-17 design to large scale production quality, which meant that a period of expansion and tooling up followed. A large number of details had to be redesigned in order to create a plane that met Army Air Corps standards. The main landing gears were replaced by fixed examples with close fitting fairings and partially faired over wheels. Engine cowlings were redesigned, streamlining of the fuselage was enhanced and the fuselage area separating the pilot's and observers' positions had become a strengthened, faired over area.

On 1 March 1935, the prototype plus 109 production aircraft were definitively or-dered for fiscal year 1935 under contract AC7326, at a cost of $ 2,047,774. Serials *35-51* to *35-160* were reserved.

To oversee the building process, the Army Air Corps sent captain Edward M. Robbins to act as liaison. After completing the modification work the plane was delivered as A-17 to the Army Air Corps on 27 July 1935, with Army Air Corps Serial *35-51*. This aircraft featured a triangularly shaped vertical tail plane. After spin recovery trials it was decided that production machines would be fitted with samples of increased size.

ONE-OFFS

During all test work Northrop had found time to build a number of projects.

First, Northrop designed and built a revised version of the unsuccessful FT-1 and FT-2 carrier-borne fighters for the Army Air Corps. This type, with all naval equipment removed, was known as the Northrop 3A. It was intended as a replacement for the Boeing P-26 fighter.

The 3A, also named the XP-948 (Northrop c/n 44), was an all-metal low wing monoplane with a retractable undercarriage fitted with a 700 hp Pratt & Whitney SR-1535-6 Twin Wasp Junior engine, driving a three-blade propeller. It featured a fully

Model 3A, or XP-948, was developed as possible successor to the Boeing P-26 (coll. N. Braas)

An early production machine, still without unit markings. This machine also lacks the perforated dive brakes.

(coll. G. Balzer)

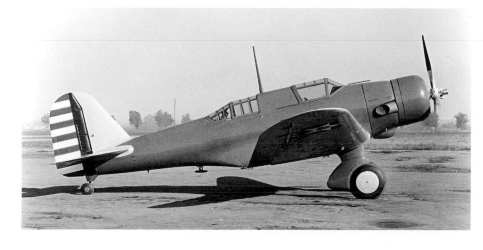

35-54, now fitted with perforated dive brakes

(coll. G. Balzer)

17 November 1935. A-17s on the assembly line. The fuselage was built in two halves. Here, the bottom half is bolted to the wing centre section. Clearly visible are the five spars instead of the more regular two main spars. The centre section held a total of six fuel tanks. In the very rear the single Model 2J2 is being assembled. (coll. G. Balzer)

Right: Another view of the factory floor, dated 7 June 1936. A number of interesting machines is visible between the regular production. At front left, one of two A-17AS. In the centre two outer wing panels are lying on threstles. These are still fitted with the criticised ailerons. Behind the wing panels, we see an A-17A, already standing on its retractable main gear legs. Behind the A-17A another interesting machine; the Model 5B. It appears to have been returned to the factory prior to delivery to its new Mexican owner. (coll. G. Balzer)

Another shot dated 7 June 1936, now taken from the opposite direction. At left is the Model 5B. The Air Corps serials are written on pieces of card and taped to the tail planes of the aircraft; 35-86 (C/N 109), 35-88 (C/N 111) and 35-89 (C/N 112) can be identified. Closest to the camera are 35-82 (C/N 105) and 35-83 (C/N 106). These were all delivered between June 18th and July 6th. (coll. G. Balzer)

A photo from the previous day, June 6th. Six shiny machines nearing completion. All aircraft have been fitted with the updated ailerons. These aircraft may include 35-74 (C/N 97) and 35-76 to 35-80, which were delivered to the Air Corps between 6 and 12 June 1936.

(coll. G. Balzer)

Two impressions of A-17s undergoing tests at Wright Field prior to delivery. (coll. G. Balzer)

ENTERING AIR CORPS SERVICE

The first production A-17 was completed on 23 December 1935, and accepted on 15 January 1936. This was actually the second production aircraft, serial *35-53*. The first production aircraft, *35-52* followed one month later. The final machine was delivered on 5 January 1937.

While first impressions were very favourable, there were teething problems as well. Some issues with propeller pitch had to be solved, and a lot of practical problems with the fixed armament had to be ironed out. After a first group of pilots assessed the A-17, complaints about the stick forces required for lateral control of the airplane had to be remedied. These were so great that pilots became physically exhausted during long flights. Low altitude flying required constant aileron control, especially in rough air - which was not uncommon so close to terra firma. The heavy aileron control was already discussed with Northrop during contract negotiations, but the company undertook no action (or was not able) to correct the problem. Northrop then offered to extend aileron area, in a similar way as was done on the Northrop 2J2 airplane. This change was introduced

enclosed cockpit. It made its first flight on during July 1935 and was shortly tested at Wright Field where A.A.C. pilots concluded that it had some stability problems and was prone to spinning. Before it could be entered in the final flight contest with competing types it crashed in the Pacific on 30 July 1935. The wreck, and its pilot, was never seen again. The fighter contest was won by Seversky with their P-35. Northrop sold the 3A design to Chance Vought Aircraft. Here, a prototype was built and flown as the V-141 and V-143 but these too never entered production.

Next, a single Gamma demonstrator (Gamma 5A) was built. It can be considered a further refined YA-13 with new cockpit design. C/n 187 was fitted with a 775 hp Wright Cyclone and carried the civil registration *X149972*, which was allocated on 4 October 1935. It was sold to the Japanese Navy as BXN1 "Northrop

Plane" in October 1935. Here it was extensively studied as an example of modern aeronautical engineering. It crashed during testing.

C/n 188 Gamma 5B was also completed during October 1935. It was a two place tandem semi-military version with sliding cockpits located on the forward fuselage and thinned down aft fuselage. The civil registration *NR14998* was issued on 2 November 1935. The plane served as a demonstrator for a possible military export order to Argentina. It received a new engine and was sold to a new Mexican owner, a certain Lt. Col. Monteiro and flew shortly wih civil registration *XA-ABI*. This may have been a cover up since it was shipped to Spain during December 1937. Here, it would be used by the Spanish Republicans during the Spanish Civil War for coastal patrols.

Technical data			
	YA-13	YA-16	A-17
Engine	Wright SR-1820F-2 Cyclone	P & W R-1830-7	P & W R-1535-11
Power	548 kW (735 hp)	950 hp	750 hp
Wingspan	48 ft (14,63 m)	48 ft	47 ft 8,5 in (14,54 m)
Length	29 ft 2 in (8,89 m)	29 ft 8 in (9,04 m)	31 ft 8 5/8 in (9,67 m)
Height	9 ft 2 in (2,79 m)	9 ft 2 in (2,79 m)	11 ft 10,5 in (3,62 m)
Empty weight	3,600 lb (1,633 kg)	-	4,874 lb (2.211 kg)
All-up weight	6,575 lb (2,982 kg)	6,700 lb (3.061 kg)	7,337 lb (3.328 kg)
Max. speed	207 mph (333 km/h) at 3,300 ft	212 mph (341 km/h)	206 mph (332 km/h) at sealevel
Range	1,100 m (1,770 km)	1,000 m (1609 km)	650 m (1046 km)
Service ceiling	21,750 ft (6,630 m)	22,000 ft (6706 m)	20,700 ft (6.310 m)
Climb rate	1,300 ft/min (396 m/min)	-	1,530 ft/min (466 m/min)
Armament	4 fixed 0.30, one flexible 0.30 mg	-	4 fixed 0.30, one flexible 0.30 mg
Bomb load	1100 lbs of bombs		20 internally stored 30 lb (13,6 kg) bombs plus four externally carried 100 lb (45 kg) bombs or flares. Two gas tanks could be fitted under the wings.

after the fifth production A-17. After implementing this, the increased aileron area had actually increased the necessary stick forces. An interim solution was found in adding the same type of balances as on the Curtiss A-12. The problems were finally tackled by altering the aileron hinge location, again in a similar way to the model 2J2. This turned out to provide acceptable aileron control, and all following production machines would be built with the modifications. During the period of modifications, eleven production machines had been delivered, and all would be retrofitted. Another critical issue had already

been addressed on the first production aircraft. After experiencing unfavourable dive qualities, the perforated dive brakes which were introduced on the BT-1 were also fitted on the A-17.

Conclusion of the reports was that all future aircraft should only be accepted with all armament fitted. This had to be tested under field conditions for at least two weeks, before any production order was to be given.

The last Gamma built by Northrop was c/n 347. It was a Gamma 2L with longer fixed undercarriage legs to accommodate a larger propeller. It was purchased by Bristol Aeroplane in the U.K. without engine. Bristol used it as an engine test-bed for the Bristol Hercules sleeve-valve radial engine. In fact it was fitted with a complete Hercules engine unit including the cowling driving a three-blade propeller. It carried the British civil registration *G-AFBT*. First flight took place at Filton in September 1937 with a 1290 hp Bristol Hercules Mk.1SM. It was used for a number of years and finally scrapped in 1946.

A-17 IN US AIR CORPS SERVICE

By Edwin Hoogschagen

35-77 at March Field, 11 June 1936. This machine shows the modified ailerons. The machine still lacks fixed machineguns, but bomb racks have been fitted. (coll. G. Balzer)

"The duty of attack aviation is to hedge-hop over the heads of the enemy, delivering a veritable hail of death with machineguns and fragmentations bombs upon the heads of troops and upon ordnance, supplies and transportation" (Flying magazine, October 1935)

As part of the 1935 plan to form a new GHQ Air Force structure, the A-17s were to be distributed over the squadrons of the 3rd and 17th Attack Groups. Each squadron was to have a war strength of 28 aircraft, but this number was never reached in practice.

The first A-17s were to be delivered to the newly formed 17th Attack Group (AG), based at March Field, Riverside California. After receiving news that this Pursuit Group (PG) would take on the attack role, a lot of energy was invested in training, studying attack manuals, following lectures on attack aviation and setting up the necessary logistics. Although still equipped with Boeing P-12 aircraft, the first months of 1935 were spent with training and studying the specifics of the new task ahead. Part of pilot training was an eight day flight to Maxwell Field, Ala. The first

A-17s were planned to be delivered during March 1936. In the meantime, P-26's were refurbished by Boeing and received flaps. This gave future A-17 pilots opportunity to train with flap equipped aircraft.

Prior to accepting their new material, the 3rd Attack Group moved from Fort Crocket, Texas, to new home base Barksdale Field, Shreveport, La. during March 1935. Unlike the 17th AG, the 3rd had a long history as Attack Group and had flown with Curtiss O-1 Falcon, Douglas O-2, Curtiss A-3 Falcon and Curtiss A-12 Shrike.

17th AG received their first A-17s during March, 1936, with the 34th receiving their first six, 73rd AS three, and Group Headquarters two machines. The aircraft were flying continuously, to allow pilots to get acquainted with their new machines. The A-17s showed impressive performance, as one pilot of 17th AG staff showed by completing a 814 mile, non-stop, flight from March Field to Hamilton Field and back in five hours, 35 minutes.

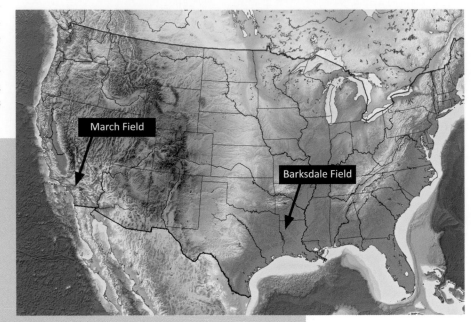

1st Wing
17th Attack Group (March Field)
    34th Attack Squadron
    73rd Attack Squadron
    95th Attack Squadron

3rd Wing
3rd Attack Group (Barksdale Field)
    8th Attack Squadron
    13th Attack Squadron
    90th Attack Squadron

Besides these units, Air Base Squadrons were created, which were responsible for supply, maintenance and service of the combat units. Air Base Squadrons (ABS) were situated at a home station and served all airdromes within a radius of approximately 200 miles. An example is the 14th Air Base Squadron (Bolling Field). This unit was formed, like many other ABS' on 1 September 1936, along with 1st and 2nd Staff Squadrons and Base Head Quarters. Commonly, a number of up to five machines flew with each of these units.

A-17s were also frequently attached, or loaned, to School Squadrons where they served with Attack flights.

35-60, which was operated with 17th AG at March Field.
(coll. G. Balzer)

Above

The centre section contained no fewer than six fuel cells, which were placed in between the wing spars. Northrop wing construction featured six wing spars instead of the more regular two main spar construction.

Left:

The machine gun armament was snugly built in to the outer wing panels and could only be reached through two small hatches. Ammo boxes had to be interchanged through these hatches as well.

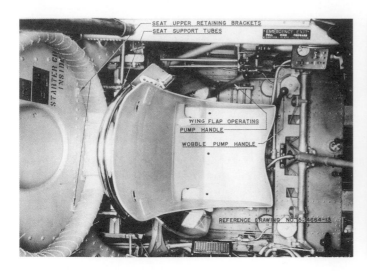

SEAT UPPER RETAINING BRACKETS
SEAT SUPPORT TUBES
EMERGENCY EXIT
WING FLAP OPERATING PUMP HANDLE
WOBBLE PUMP HANDLE
REFERENCE DRAWING NO 5-4664-13

465489 CASE (REF)
MAPS
FLIGHT REPORT
WIRING DIAGRAM
INSTRUCTION MANUAL
REFERENCE DWG. NO. 576516

Overview of the pilot's cockpit. As the drawing reveals, a second set of controls could be fitted in the rear cockpit

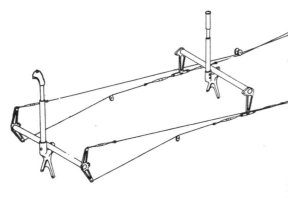

33 A1755-TYPE-BC-AE-232 CONTROL BOX
33A1757-TYPE-MC-125 TUNING CONTROL
33 A1756-TYPE-BC-AE-231 CONTROL BOX
277287 CONTROL BOX
477411 INTERPHONE CONTROL BOX
FLARE RELEASE
REFERENCE DRAWING NO. 476514

467550 GUN SWITCH BOX
IGNITION SWITCH
476030 SWITCH BOX
FLAP WARNING
REFERENCE DWG. NO. 575892

567728 LIFE RAFT
CONTAINER

578282 BAGGAGE
COMPARTMENT

REFERENCE DRAWING
NO. 564664

BEARING PIN ATTACHING BOLT

165820 PIN

562534 PEDAL ASSEMBLY

178755 SPRING
265845 SOCKET ASSEMBLY
069534 WIND VANE SIGHT TYPE D-4
0150958 MOUNT GUN SIGHT TYPE C-1

0150973 RING SIGHT
TYPE B-8

30-616 TYPE L-4 AMM.
MAGAZINES
BROWNING M-1 MACHINE GUN

A look inside the rear cockpit shows various details such as installation of the radio, gun ring and stored ammunition boxes. The observers' gun could be stowed in a recess which was closed with a pair of hatches.

35D1209 TYPE BC-AE-198 ANT. SWITCH RELAY
34B5892-TYPE-C-168 COIL
34B5892-TYPE-C-169 COIL
34B5892-TYPE-C-172 COIL
35A1219-TYPE C-181
COIL SET

33A1754-TYPE BC-AE-229 RECEIVER
TYPE C-153 COIL

33A1753-TYPE BC-AE-230 TRANSMITTER
TYPE C-180 COIL SET

REFERENCE DWG
NO. 561521

RECEIVER

GROUND WIRES

A 3rd AG machine. During the first period of flying, no unit badges were painted on the aircraft. 3rd AG machines carried thin registries at the top of the tail plane, while 17th AG carried fat registries in the middle of the tail plane. There was no common system for the application of registries until well into 1937. (coll. E. Hooyschagen)

While A-17s were still being delivered to the Air Corps, a fatal flying accident took place at Wright Field. Captain William L. Scott Jr., of the Air Corps Engineering School, was killed on 17 May when *35-52* crashed two miles east of the airfield when it entered a spin from which it could not be recovered. *35-52* was the second built A-17, and was delivered on 13 February.

One of 3rd AG's new A-17s was first displayed to the public at Hamilton Field during the National Defense Week Demonstration held there on 29, 30 and 31st May. The 3rd AG continued to fly their Curtiss A-12s alongside their new

Northrops. During June, 13th and 90th AS performed Army cooperation missions. First, 90th AS sent three A-12s to Maxwell Field, Ala. for a simulated attack on ground forces near Fort Benning using machine gun and mustard gas attacks (the planes were fitted with two 18 gallon tanks which were loaded with lime water instead of the actual mustard gas). On 20 June 13th AS sent three A-17s to Fort Sill, Okla., where they were joined by three O-43 planes from Brook Field to form "Red Air Force". Their task during the exercise was to repulse a ground attack by "Blue Force". The exercise started with machine gun and bombing attacks on small

groups of troops on 22 June. The simulated gas attacks were hampered by high winds, but otherwise the mission was declared a success.

The exercise was ended on the 26th. The 3rd AG was again out in the field to participate in the manoeuvres of the Second Army. Sixteen machines left Barksdale Field on 1 August, destination Selfridge Field, Michigan for a nine day exercise.

Six machines from 34th AS, 7 July 1936.
(coll. G. Balzer)

After receiving the A-17 order, Northrop built Gamma model 2J2 (c/n 186) which was in fact a demilitarized A-17. It was completed before deliveries of A-17 had started, on 18 December 1935. Power was provided by a 550 hp P&W R-1340 Wasp S3 radial engine driving a 3-blade propeller. After the earlier experimental semi retracting landing gear of the 2F the Northrop design team created a more advanced variant with inward retracting landing gear legs. The main wheels fitted in a recessment in the leading edge of the wing, with the wheel partly protruding forward. The machine was offered for testing on 10 January 1936 and it was entered in the pre-war competition for a new advanced trainer. It lost from the competing North American design later known as the T-6. Afterwards, it was used as a company hack by Northrop,

and later for some years as a company plane by Douglas. It was then sold to civilian owners until it crashed on 19 May 1945.

A type closely related to the Gamma model 2J2 was Gamma 5D c/n 291. It was fitted a with a 550 hp Pratt &Whitney S3H-1 radial engine. It received registration X16091 on 15 September 1936 but was eventually sold to Japan in an undocumented transaction. In Japan it was known as BXN2 "Northrop Plane" and was used by Nakajima and may have been of influence for the design and construction of the later Type 97 'Kate' attack bomber. In 1939 it must have been used for a short period for aerial reconnaissance (espionage flights!) above China and Russia by Manchurian Airlines.

Model 2J2 was an unarmed variant of the A-17, fitted with retractable undercarriage, modified cockpit hoods and two blade propeller. (coll. G. Balzer)

While A-17 production was still under way, Northrop sent a proposal to the Army Air Corps to develop a version of the A-17 with a retractable landing gear similar to that of model 2J2. Although empty weight of the airplane would increase an estimated 230 lbs (104 kg), Northrop reasoned that the improved machine would have improved range and speed, with an increased maximum bomb load of 1200 lbs (544 kg). The air corps placed an order for 100 machines, plus spare parts equivalent to 15 airplanes, at a total cost of $ 2,560,074.00. The machines would enter service as A-17A. Serials 36-162 to 36-261 were reserved.

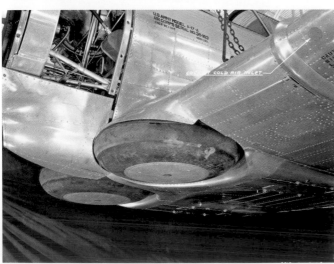

The Pratt & Whitney R-1535 Twin Wasp Junior engine was a widely used 14 cylinder air cooled radial engine. It was introduced in 1932 as development from the R-985 9 cylinder radial and had a displacement of 1535 In3. The engine ran on 87 Octane fuel. Two variants of the R-1535 were used in the A-17, the R-1535-13 offered a slightly higher RPM resulting in a higher power output. The engine was produced in various subtypes until 1941, with 2880 built.

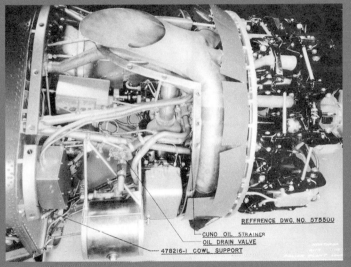

	R-1535-11 (Northrop A-17)	R-1535-13 (Northrop A-17A)
Compression Ratio	6.7:1	6.7:1
Sea level Speed R.P.M.	2500	2500
Normal Brake HP at Sea level	750	735
Rated Altitude	2500	2500
Altitude Speed R.P.M.	2500	2500
Normal Brake HP at Rated Altitude	725	750
Maximum Brake HP at 2625 RPM		825
Weight	1102 lb	1110 lb

Although the A-17AS did not carry any armament, it was still equipped with standard A-17 styled perforated flaps.

(coll. G. Balzer)

VIP variant

Before the series production of the A-17As started, two airframes received special attention. They were similar to the A-17A, but were intended as high speed unarmed staff transport and were able to carry a pilot and two passengers. Both planes were recognizable by the modified canopy and engine cover of slightly larger diameter.

They were powered by a 600 hp direct-drive Pratt & Whitney R-1340-41 nine cylinder radial driving a two blade propeller. Type designation was A-17AS. The first was c/n 289 with military registry *36-349*. It was assigned as personal aircraft of the Chief of the Air Corps, major-general Oscar Westover and was delivered on 17 July 1936.

Right:

36-349 is being handed over to general Westover.

(coll. G. Balzer)

The second A-17AS was c/n 290 (*36-350*) and was delivered a few days ahead of *36-349* on 12 July 1936. It was assigned to brigadier-general Henry H. Arnold.

Both aircraft were adorned with the 14th BG insignia, the Capital Dome. 36-349 was numbered "1" and received two stars on the tail plane. 36-350 was numbered "2" and had one star on the tail plane. Both aircraft were used on inspection trips throughout the United States and were normally piloted by Westover and Arnold themselves. General Westover picked up his new personal aircraft while on a month long tour through the country and got well acquainted with it for the second half of his journey.

*General Arnold and mr. Northrop
in front of 36-350.*
(coll. G. Balzer)

The A-17AS featured a baggage compartment directly behind the rear cockpit.
(coll. G. Balzer)

Right: This photo shows that the A-17AS originally had the option to be operated as three seater.
(coll. G. Balzer)

36-349 in its livery featuring the capitol dome and two stars on the rudder.
(coll. G. Balzer)

36-350 carried a single star on its rudder.
(coll. G. Balzer)

Technical data A-17 A	
Engine	P & W R-1535-13
Power	825 hp
Wingspan	47 ft 9 in (14,55 m)
Length	31 ft 8 in (9,65 m)
Height	12 ft (3,66 m)
Empty weight	5,106 lb (2.316 kg)
All-up weight	7,337 lb (3.328 kg)
Max. speed	206 m/ph (354 km/h) at sea level
Range	730 m (1.175 km) maximum: 1,195 m (1.923 km)
Service ceiling	19,400 ft (5.915 m)
Climb rate	1,350 ft/min (411 m/min)
Armament	4 fixed 0.30, one flexible 0.30 mg
Bombload	20 internally stored 30 lb (13,6 kg bombs plus four externally carried 100 lb (45 kg) bombs or flares. Two gas tanks could be fitted under the wings.

The very first A-17A suffered from many teething problems. Series production could commence after tackling technical, but also labour trouble.

(coll. G. Balzer)

Middle:
Anonymous machine, undergoing trials at Wright Field.

(coll. G. Balzer)

A-17 deliveries were completed between April and December 1937. (coll. G. Balzer)

Technical issues with the retractable landing gear slowed down production of the rest of the machines. *36-162* was involved in two accidents caused by landing gear failures. It was finally delivered on 4 February 1937. Personnel strikes also prevented timely delivery of the remaining aircraft. Continuous labour problems led to the aircraft being refused by the War Department, which threatened to only take over the planes after the problems had been solved. Unions were forming among car and aircraft industry labourers, demanding higher wages. After a sit down strike on 27 February, the situation escalated quickly. After such a strike at the Douglas company, Northrop employees followed their example. Douglas responded by acquiring the remaining 49 per cent of Northrop shares on 5 April 1937 in an attempt to disrupt the rout. Work resumed and delivery of the A-17A series finally took place between April and December 1937.

In the meantime, yet another strike led to the dissolvement of the Northrop Corporation which then became the El Segundo Division of Douglas.

Jack Northrop resigned and left Douglas on 1 January 1938 and established his own aircraft manufacturing company: Northrop Aircraft, inc. It was a major independent United States aircraft manufacturer until it merged with Grumman in 1994 to form Northrop Grumman.

Jack Northrop hoped to build his dream: the Flying Wing. This finally resulted in the XB-35 and YB-49 flying wing bombers but these were built in small numbers only and were never operational! The more successful types of Northrop until the eighties were far from flying wings!

Jack Northrop died at 85 on 18 February 1981, but before he died he was present at his company at the start of the B-2 Spirit flying wing stealth bomber project. Unfortunately he never saw the first flight of his flying wing dream!

US East Coast.

Shortly after delivery of the two A-17AS' the first production A-17A (c/n 189, *36-162*) made its first flight on 16 July 1936 and was delivered to the air corps on 12 August.

37th AS, based at Langley Field, Virginia, which operated the last Curtiss A-8s traded their machines for a group of eight factory fresh A-17s during early August. 37th AS was attached to 8th PG, which on its turn was part of 2nd Wing. Apparently, a number of officers were unimpressed by the qualities of their new planes. The Air Force News Letter, in a short news article dated 1 December 1936, reported rather cunningly after studying a filed report with complaints, that these officers may have been better off with a flock of their trusty A-8s. Their complaints about some of the equipment aboard the A-17 were described as minor deficiencies. The mentioned equipment would be missed as a luxury compared to the A-8!

On 21 August, a formation of five A-17s from 73rd AS made a successful non-stop trip from March Field to Kelly Field, San Antonio, Tex. with an average speed of 143.4 miles per hour. The flight was made to test fuel and oil consumption of the pla-

A 37th AS machine sporting the unit badge, a charging lion on a white spade with black border. The engine cowling in white. (coll. G. Balzer)

nes. Kelly Field was reached after a flight with mostly head winds but the planes still had an average fuel load of 50 gallons left after landing. No mechanical trouble developed and the return flight was started with only servicing and the routine checks of engines.

An 18 plane strong flight of A-17s was presented to the public at the 1936 National Air Races at Los Angeles, held 4, 5 and 6 September. Eight of these planes left Barksdale Field on 1 September and took the aviators via Midland, Tex., El Paso, Tex., March Field, Calif. to Mines Field, California. While at March Field, formation flying was practiced, in particular a formation forming the letters "LA" which would be displayed on Sunday. The Air Corps new attack planes had to settle for a humble role, as they also had to act as target for a formation of 18 Boeing P-26's from the 79th Pursuit Squadron (PS). Although, of course, none of the interceptions caused any casualties, one of the A-17s was damaged during the event, as Steve Wittman's racing plane "Chief Oshkosh" ran into one of the parked planes after its engine had cut out on Saturday. Still, the planned activities continued on Sunday, but heavy fog prevented the men to display their "LA" formation to the public. The return trip to Barksdale Field commenced on 9 September. After several stops to take in fuel and a nights' rest at Hensley Field, Tex., all aircraft arrived safely at Barksdale Field. One plane lagged behind due minor technical trouble, but on the whole, the 168'45 hour flight was a complete success. Less than a week later, on 11 September, 17th AG was confronted with the loss of an A-17 and its crew. 35-75, attached to HQ squadron crashed 1.5 miles west of March Field. Pilot 2nd lt. James T. Carter Jr. and private Milton J. Cutting were killed. The plane had been delivered just over three months earlier, on 3 June, and had logged 208h 55m.

During the same month, 90th AS was again active on more serious business. Twelve aircraft were participating in a field exercise during September, this time taking them to Fort Crocket, Texas, for a week long exercise. Departure was on the 7th. All materials, spare parts and necessary equipment were brought to the destination by motor vehicles. The aircraft had to operate under primitive conditions, some 100 miles from the theoretical front. With excellent weather, two missions could be flown every day. Strafing attacks, bombing with 17 pound shrapnel and 100 pound high explosive bombs, and mustard gas attacks were all part of the drill. After 205 trouble-free flight hours the detachment returned to Barksdale Field on the 12th. 8th AS made a similar trip to Fort Crocket from 21 to 26 September and completed their field exercises without mishaps.

A mass take off of no less than 18 aircraft for the National Air Races at Los Angeles (coll. E. Hoogschagen)

90th AS sent out its planes for a demonstration at Sherman Field, Kansas on the 25th. Different scenarios were displayed, such as a mustard gas attack on a marching column, low altitude, high speed, bombing of an infantry battalion with 30 pound bombs and repelling an interception by patrolling pursuit aircraft. Acting as a dummy target would be a recurring task for the 3rd AG, as their aircraft acted as enemy forces for the 20th PG, also stationed at Barksdale Field.

Twelve 8th AS A-17s flew to Natchitoches, Louisana, for the 1936-37 field exercises which took place from 5 to 10 October. The squadron sent out a similar number of aircraft to the 3rd wing manoeuvres at New Orleans, which started on 5 November.

Large formations of A-17s were a frequent sight across the USA. Tail number 76 was an 37th AS machine, which crashed on 25 March 1937. (coll S. Donacik)

Left: *3rd AG planes at their home base of Barksdale Field, which was opened early 1933. (coll. G. Balzer)*

Below: *The seven A-17s and two P-12s from 34th AS prior to their departure to Bakersfield. March Field, 20 October 1936.*

(coll. G. Balzer)

A large 17th AG contingent was active during large scale manoeuvres held at Bakersfield, Calif. from 19 October until 18 November.

It was their first major exercise since receiving their new equipment. The exercises were flown together with detachments of 19th BG. Each of the 17th AG's squadrons would be active during one week. 34th AS would bear the brunt of the first week and sent out seven A-17s and two Boeing P-12 on 21 October and returned the 28th. 73rd AS, leaving on 28th, sent a five plane formation, while the 95th AS concluded the 17th AG's attendance starting 11 November. In all, 24 aircraft, including some of the AG Headquarters Squadron, would participate. No less than 2500 50 pound bombs were brought to Bakers-

All kinds of mishaps		
During the first months of flying on the A-17 several incidents occurred.		
13 jun 1936	*35-60*	13th AS - landing accident at Barksdale Field
7 jul 1936	*35-57*	13th AS - Take off accident at Barksdale Field
21 jul 1936	*35-88*	73rd AS - Ground looped during landing, March Field, possibly written off
3 aug 1936	*35-91*	95th AS - damaged during landing, Saugus Calif.
11 sept 1936	*35-75*	17th AG HQSQ - crashed in bad weather, written off
11 sept 1936	*35-74*	? - unknown accident, possibly written off
15 sept 1936	*35-92*	95th AS - crashed during landing, Oceanside Calif.
16 Oct 1936	*35-65* and *35-67*	13th AS. were badly damaged in a ground collision at Barksdale Field. The planes had been in service for just five months. Both aircraft could be repaired.
12 Nov 1936	*35-147*	Forced landing during delivery flight, 5 m. south west of Corona, Calif.
04 Dec 1936	*35-143*	4ABS - Forced landing following engine trouble, Bakersfield Calif.

An unarmed 17th AG machine visiting Randolph Field, 23 January 1937. (coll. G. Balzer)

field. Main activity was practicing the art of glide bombing at the Muroc Bombing and Gunnery Range, Muroc, Calif. During a glide bombing run the planes would be coming down in a steep glide of about 35° to a maximum of 55° during their attack run, dropping their ordnance at an altitude of 1000 ft or more.

1937 started tragically when lt. col. Frederick I. Eglin[1] and his passenger lt. Howard E. Shelton Jr. (US Navy) were killed in a flying accident on 1 January 1937. They were en route from Langley Field to Max-well Field. The men may have been surprised by poor weather and heavy fog over the Appalachian peaks. The wreckage of their A-17, *35-97*, was found near the top of Cheaha mountain.

---

1   The former Valparaiso Bombing and Gunnery Base was renamed Eglin Air Force Base in honor of F.I. Eglin.

Colours and markings

The A-17s were delivered in the standard Air Corps paint scheme which was first advised in January 1934 and implemented in specification No. 98-24113, dated 23 May 1934. Standardisation of the paint scheme was needed to prevent large stocks of various paints. For the A-17 this meant that the whole of the fuselage, landing gear struts and wheel spats were painted in Light Blue No. 23 and the wings and tail surfaces in Yellow No. 4. A flat black walkway was painted on the wing root.

Nationality markings were standardised in 1926 and included star insignia on the top and bottom of the wings, while the tail rudder was divided in a dark blue vertical bar, 1/3 wide, and thirteen alternate white (six) and red (seven) stripes. The inscription U.S. ARMY was stencilled in black on the outer wing panels.

At the moment of delivery a minimum of service stencilling was placed. The aircraft type, Air Corps Serial number and crew weight was stencilled on the port side of the fuselage, and the location of a fire extinguisher was indicated in a red square. More service markings were added once the machines reached the squadrons.

When the A-17As were built the paint specifications had already changed. The introduction of Alclad in aircraft production vastly improved protection from corrosion, which resulted in the introduction of unpainted natural metal aircraft finishes. Weight of the aircraft decreased slightly, but the main advantage was that maintenance and overhaul was less time consuming and at lower costs. The areas covered with linen were treated with aluminium dope and a Flat Bronze Green No. 9 anti-glare panel in front of the wind screen was introduced.

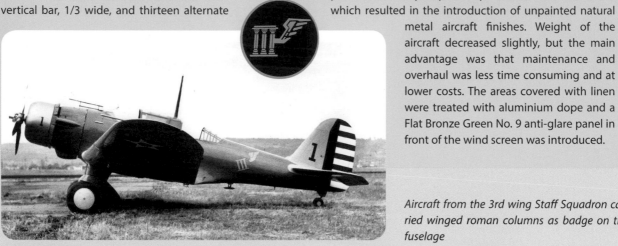

Aircraft from the 3rd wing Staff Squadron carried winged roman columns as badge on the fuselage

1st Wing

17 AG   34th AS
*   73rd AS*
*   95th AS*

34th AS   73rd AS   95th AS

2nd Wing

8th PG   37th AS

37th AS

3rd Wing

3rd AG   8th AS
*   13th AS*
*   90th AS*

8th AS   13th AS   90th AS

The wheel spats on this 37th AS machine have been removed. These spats were often removed to prevent mud and debris building up inside.
(coll. G. Balzer)

Center: 8th AS planes with the E added to the plane numbers.
(coll. G. Balzer)

37th AS was involved in a number of winter trials during January 1937. Earlier, a representative of Eclipse Aviation Corporation visited the San Antonio Air Depot, for a conference on the cold weather procedures for the type C-20 starters which were fitted on the A-17s. Langley Field had been in a very poor state after adverse weather. Parts of the field were completely water logged. Conditions finally improved, and flying was finally possible again on the 25th. One of the 37th AS planes was flown to Duncan Field, where an oil dilution system was installed. This system was necessary for cold weather engine starting and flying. A Type E-4 Radio Compass was successfully tested in another machine. Other tests included the experimental air heater which was fitted in the cockpit of one of the machines. Cold weather flying

gave particular challenges, such as ice forming up in the carburettor inlet, ice and mud building up in the wheel pants and ice and debris being launched at the tail planes during take-off and landing. Winter trials were also held at Selfridge Field. Here, 8th AS had settled down on 2 February and was actively flying 19 A-17s in a cold weather provisional test group. This manoeuvre consisted of formation flying, night flying, testing of service equipment, machine gunnery and bombing. Other practical issues were looked into. Clothing for both flying and technical personnel were tested. The program was concluded on 24 February.

The Harmon Efficiency Trophy was awarded to the 90th AS on 13 March. It was honoured for having the highest efficiency

rating. 90th AS gained the highest score, after considering the number of flight hours, the average numbers of aircraft available and the number of forced landings during the calendar year. The trophy was presented to the commanding officer Captain William N. Amis. 8th and 13th AS, followed by the pursuiters of 20th PG, flew a review flight after the presentation. Regular training continued for 90th AS, when six planes were sent to Fort Benning on 1 April for chemical warfare training. On 21 November another prize, the Frank Luke Trophy, was awarded to the 3rd AG as a whole for having achieved the lowest accident rate per thousand flying hours.

A 13th AS formation. The aircraft carry an E above the individual plane number. This indicated that the 13th AS had won the Harmon trophy and was allowed to carry this E until the next edition of the trophy. (Coll. E. Hoogschagen)

37th AS was not only testing its equipment in cold conditions. During the 2nd Wing Field Exercise, the squadron had set up camp in Rocky Mount, N.C. During the week long exercise, in similar freezing conditions as the earlier winter trials, the squadron gained valuable experience. The otherwise successful week was marred by an incident on 16 April, when an A-17 had struck an ambulance vehicle while taxiing. The right wing was damaged and had to be replaced. 37th AS took part in demonstrations for vice president John Garner who visited Langley Field on 24 April.

An 8th AS machine at Griffith Park, April 1937. (coll. G. Balzer)

Left: An 13th AS machine, with the squadron badge applied, a skeleton wielding a scythe over a black background. (coll. G. Balzer)

The month of May 1937 saw a large scale concentration of Army Air Corps aircraft. No less than 244 aircraft gathered at March Field and Hamilton Field, California, for the West Coast GHQ Manoeuvres. A number of commercial air fields were used as satellite fields. Goal of the manoeuvres was to fly a large number of aircraft of different types, from different units in full-strength. It was tried to operate the aircraft under combat like conditions from a number of airfields in a confined geographical area. The 3rd AG participated as the "Attacking Force" and its squadrons were dispersed over the following fields;

Attacking Force		
HG Squadron	5 A-17	Bakersfield
13th AS	28 A-17	Delano (including 7 plane detachment of 37th AS)
34th AS	28 A-17	Visalia
90th AS	28 A-17	Bakersfield

A-17s and B-10s lined up at March Field to take part in the war games at Muroc Lake. (Coll. E. Hoogschagen)

36-170 was among the first aircraft which were delivered during April 1937. (coll. G. Balzer)

Most of the bombing and gas attack exercises were flown over Muroc Lake, where targets were lined out in chalk. Air defences were erected at Muroc Lake. Anti-Aircraft guns and search lights were concentrated around the town of Muroc. Personnel of the 1st PG had set up camp there, while their P-26 fighters operated from the lake bed. Finally, a grid of air observer posts was set up around the theatre. 13th AS had been assigned with chemical missions, which consisted of contaminating the Pursuit airdrome and the anti-aircraft emplacements. 37th AS, which had just received its chemical equipment in April, joined 13th AS to gain experience in chemical operations.

Valuable lessons had been learned, not only from the flying itself, but also other practical issues, such as logistics.

When the pilots of 90th AS returned to Barksdale Field, they were surprised to find six brand new A-17A airplanes waiting for them. The pilots quickly took to the air and were checked off, after being instructed in the use of the biggest novelty, the retractable landing gear. In a short time span the number of aircraft had risen to twelve, and the training scheme was expanded with night flying.

A 90th AS machine while visiting Chevalier Field, NAS Pensacola. (coll. G. Balzer)

The 37th AS, meanwhile, was once again in the field for gunnery exercises. Six aircraft, accompanied by 31 Consolidated PB-2As from 36th PS and one Glenn Martin B-10B left Langley Field on 24 June, destination Virginia Beach. Here, a program of aerial gunnery and ground attacks was carried out. Another part of the exercise was a navigation flight, involving twelve PB-2As and six A-17s. The group left on 28 June and reached Buffalo N.Y., via Patterson Field and Langley Field. The following day the formation reached Wright Field. Here, the Curtiss factories and the Material Division were visited. The final leg of the flight was undertaken at night. Take off from Wright Field was at 02:00 and the aviators returned to Virginia Beach four hours later, for the remainder of the gunnery exercise, which would last up to 8 July.

By September 1937, the number of new A-17As at hand had increased significantly. This meant that a large number of A-17 aircraft were redeployed to School Squadrons.

This process had already started in the first months of 1937, and took more momentum after introduction of A-17As at first line units. Large amounts of A-17s were used by 63rd and 64th School Squadron (SchS) at Kelly Field. A number of A-17s served with 91st SchS, based at Maxwell Field.

This squadron was activated on 1 September 1936. A wide range of aircraft types was used by the 91st SchS, such as P-12, O-19, O-46, B-6, A-12. At least five A-17s were used at Maxwell Field. A-17s flew at a number of other School Squadrons - the 1st, based at Chanute Field, 33rd, based at Lowrey Field, 52nd, based at Randolph Field and the Air Corps Technical School, based at Langley Field (later transferred to Chanute Field).

Top and centre:
17th AG aircraft lined up, 9th August 1937.
(coll. G. Balzer)

Students in front of an A-17 at Kelly Field.
(coll. G. Balzer)

During October 1937 the A-17-AS were fitted with a geared R-1340-45 three blade propeller and deeper engine cowling. A black anti-glare panel was painted on the top of the engine cowling.

Commonly, aircraft were being rotated from first line units, served with a SchS before being redeployed to other units.

On 10 September 21 A-17As of the 17th AG took off on an extended unit navigation flight. Stops were made at Albuquerque (New Mexico), Fort Leavenworth (Kansas), Selfridge Field (Michigan), Patterson Field (Ohio), Maxwell Field (Alabama), Barksdale Field (Lousiana), El Paso (Texas). All returned to March Field on the 18th. The entire flight was so arranged as to provide training in the various types of problems required of attack aviation. Rendezvous problems as well as attack and navigation missions were flown. Another purpose of the flight was to familiarize personnel on the west coast with the facilities and operation of other air bases throughout the country.

37th AS performed an experimental instrument flight on 5 November. Three aircraft, flying at different altitudes, were radio guided to Middletown, Pa. by a fourth plane, which performed navigating and directing. The crew of the guiding plane was able to give commands by radio and all four aircraft successfully reached their objective of Middletown.

The 3rd AG was able to send no less than 36 machines from 13th and 90th AS on a parade flight over New York on 21 September.

An A-17 carrying the Air Corps Technical School badge. This was also used by 10th ABS and 1st SchS.
(coll. G. Balzer)

Below: 38-327 was the first example of the second series of 29 A-17As. It was delivered on 10 June 1938.
(coll. G. Balzer)

On 9 November a contract was signed for delivery of 29 additional A-17As at a total cost of $654,155.90. A further 43 Pratt & Whitney 1535-13 geared radial engines plus spare parts were purchased at a total cost of $447,328.80 on 2 December 1937. Delivery of this second batch started on 10 June 1938 and was finished on 31 August 1938.

✪ 35-133, attached to 73rd AS, crashed on 15 January at Clyde Ranch, California. 2nd lt. Hudson H. Upham and his observer survived the accident.

✪ 73rd AS lost another machine on 27 March, when 35-135 crashed into the southern slope op Mount McKinley. The accident claimed the life of 2nd lt. Robert C. Love and his passenger, pvt Emory J. Parsons.

✪ Two days earlier, 25th of March, 35-148 from 37th AS was lost after engine failure at Messick, Virginia.

✪ 35-53, attached to 1st SchS from Chanute Field was lost following engine failure during the morning of 6 May 1937 when it crashed 2 miles south east of Paris, Illinois. Pilot 1st lt. Berkeley E. Nelson escaped unhurt.

✪ On 26 June 35-156, an A-17 was lost in a crash near Timmonsville, South Carolina. Pilot Marion Huggins was killed in the incident.

✪ 35-120 was involved in a mid-air collision on 24 July 1937 over Range B, Barksdale Field.

✪ 37th AS lost a machine on 1 August, when 35-104 crashed after taking off from Edgewood Arsenal. It was piloted by lt. William R. Enyart, a trainee pilot attached to the 37th AS. Enyart, and passenger Major Guy H. Moates, were killed instantly.

✪ Another accident occurred on 1 August, involving 35-103. It was possibly written off.

✪ 35-105 was lost with its crew on 18 August 1937. It was flown by col. William C. McChord, who was chief of the training and Operations Division.

✪ An A-17A from 13th AS was lost in a fatal crash on 1 December 1937. The crew, 2nd Lt. Charles W. Field and Private George W. Bolton jr. lost their lives when their plane, 36-210 went into a tail spin and crashed 2,5 miles southeast of Avery, Texas.

✪ *An A-17 attached to the Attack Section of the Air Corps Advanced Flying School (63rd School Squadron), Kelly Field, crashed in poor weather on 7 December. The machine, 35-68 was flown by Mexican student Capitan F.A. Avelino while he was taking part in a student flight with 16 other students. While the formation was taking off from Hensley Field, the weather was fine, but during departure with five minute take off intervals, deteriorating weather conditions were being reported. Low cloud was encountered on the way to Kelly Field. It was decided that the flight had to be aborted and instructions to return to Hensley Field were sent out with radio broadcast. Sixteen of the pilots carried out these instructions, but unfortunately, Capitan Avelino did not return. His plane had been heard circling over Austin at low altitude, before crashing into a house. He, and a two year old child, were killed instantly.*

Following the discovery of cracks in the upper skinning on the right wing of an A-17A, 55 A-17 and A-17A aircraft were inspected at Barksdale Field in the period 7 to 13 January 1938. The inspection revealed that 42 out of 50 A-17As had cracks in the right wing, varying from one to twelve cracks. Three of the five inspected A-17s also had cracks in the upper right wing skin. It was obvious that these cracks were caused by fatigue, and would continue to develop as flight hours increased. Two aircraft inspected on 8 January were found to have new cracks when they returned from cross country trips on 10 January, after approximately eight additional flying hours. Placing patches over cracks was considered to be no option. Not only would the added external patching disrupt air flow, it was also quickly discovered that cracks continued to develop after a wing had been repaired this way. The best remedy was to remove all affected wing skinning, add reinforcements to a number of places and reskin the area. This repair project started after the Northrop company sent drawings and instructions. When aircraft were sent to the San Antonio Air Depot, the horizontal tail planes were repaired in a similar fashion. In addition to the reworking of surfaces, throat microphone amplifiers, interphone amplifiers and carburettor mixture thermometers were installed. By 9 June 30 aircraft had been modified, and work continued at a pace of three aircraft per week.

Metal fatigue lowered the number of airworthy aircraft drastically during the first half of 1938. This 90th AS machine was photographed when refurbished aircraft were becoming available again. The tailnumber is repeated on the port wing, size and location were dictated by the application of the U.S. ARMY and air corps roundel. (coll. G. Balzer)

73rd AS planes sharing the aprons of March Field with a number of Y1B-17 aircraft. (coll. G. Balzer)

Another modification, which would be fitted during the second half of 1938 was an exhaust silencer.

The Air Corps loaned an A-17A, from 90th AS, to NACA for this purpose. An exhaust silencer decreased noise on the ground, and eliminated exhaust glare during night flying. Several variants of exhaust silencers were studied by NACA at Langley Field.

37th AS was dissolved on 31 January 1938. During February four of their remaining five aircraft were transferred to other units. Two aircraft left for Bolling Field, while two others were loaned to the 8th PG, where the 8th PS found themselves without aircraft after their PB-2A's had been grounded. Earlier, one of 37th AS' planes had been involved in a mid-air collision with a bird, causing considerable da-

mage to the leading edge of the right horizontal stabilizer.

17th AG was ordered to assist after a flooding disaster devastated large areas in Southern California. Airplanes constantly flew over the flooded areas, directing and facilitating rescue work by means of radio communication. In three days' time, from 3 to 5 March, 37 missions were flown, to-

An 90th AS machine, with tail code AC-85, was sent to NACA to receive an experimental exhaust muffler. (coll. G. Balzer)

A close up view on the muffler on 36-333.
(coll. G. Balzer)

Centre:
A 3rd wing staff plane, with WC registry.
(coll. G. Balzer)

talling 51"40 flying hours. 14 sorties were flown in order to locate victims, four to locate washed-out roads, eight to locate destroyed bridges, six to check the condition of dams, two to check broken power lines and two flights were made to drop food to isolated people. One liaison flight was made to Los Angeles.

A large part of May was spent with a large scale exercise, which involved 131 aircraft. Starting 1 May, aircraft from all over the United Stated concentrated to the north eastern states. During the period 12 to 17 May combined manoeuvres were flown.

May 1938		
1st Wing		
	HQ:	5 A-17s, 1 B-18 and 1 C-36 to New Haven, Connecticut
	17th AG HQ:	5 A-17As to Groton, Connecticut
	95th AS:	10 A-17As to Rentschler Field, Hartford Connecticut
2nd Wing		
	HQ:	4 A-17s, 1 B-18 and 1 C-36 to Middletown, Pennsilvania
	3rd AG HQ:	5 A-17As to Camden, New Jersey
	13th AS:	10 A-17As to Allentown Pensilvania
3rd Wing		
	HQ:	5 A-17s to Mitchell Field, New York.

Top and bottom

The application of experimental camouflage to the aircraft of 13th and 95th AS is of interest. The aircraft received a coat of water based paint in dark green, olive drab and neutral grey/or sea green. (coll. G. Balzer)

Middle:

Impressions of the application of a water soluble experimental camouflage paint scheme on an A-17A. (Profile by Alexandre Guedes)

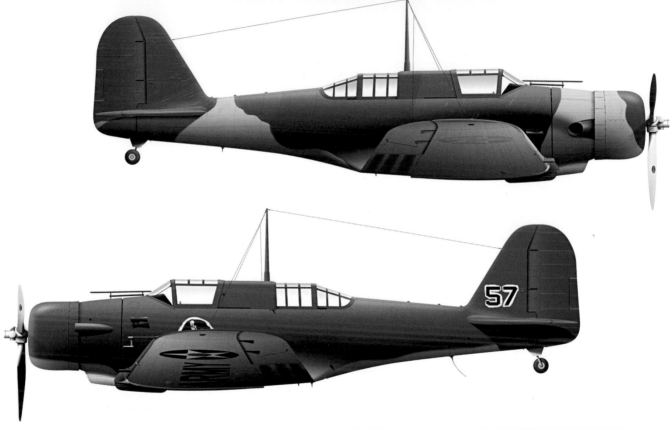

A single A-17 returned to the 1938 National Air Races held at Cleveland, Ohio.

In stark contrast to the 1936 event, the A-17 was part of a small static display of aircraft. Star of the show was 27th PS, showing off their brand new Seversky P-35 fighter aircraft. The A-17s low profile attendance may illustrate the position of the single engined attack aircraft during late 1938.

SHAPE OF THINGS TO COME

Starting July 1937, Curtiss delivered 13 twin engine Curtiss A-18's, which were ordered on 23 July 1936. All machines were delivered to the 8th AS which received instruction to compare the twin engine attacker with the A-17.

Following the small series of Curtiss A-18 twin engined attackers, a new generation of light bombers had been developed. Of these, the Douglas A-20 (or DB-7B) would play a major role in the future of attack aviation. This Douglas product took to the air for the first time on 26 October 1938.

Despite being slowly overtaken by aeronautical developments, the A-17s continued to show their worth as dependable work horses. Lt. Col. Ralph H. Hooton chose one as his personal aircraft when he was appointed as air attaché to Bolivia and Chile, departing Washington D.C and flying his aircraft via Brownsvile, Texas, Albrook Field, Guatemala City, Cali (Colombia), Guayaquil (Ecuador), Talara, Lima (Peru) to Antofagasta, Chile.

3rd AG had a busy month during October. 25 A-17A's of 13th AS departed from Barksdale Field on 1 October. They were to join an anti-aircraft manoeuvre at Pope Field, which took place from 3 to 17 October. Large amounts of anti-aircraft artillery was concentrated in the area of the field. 13th AS was operating together with bomb squadrons, operating B-10, B-17 and B-18 types and a reconnaissance squadron, flying B-10s. The attack aircraft conducted various attack missions during the break of light and at night. Independent attack

missions were flown, other flights were carried out in support of bombers. On one particular attack flight, twelve A-17s were able to fly over the field without opposition, but instead of attacking the artillery, they layed a smoke screen in order to obscure the following attacking bombers from sight. According to a witnessing coast guard artillery officer, who flew along with one of the attack aircraft, the effect of this smoke screen was questionable. Unfavourable weather conditions were welcomed, and they were used to accustom flyers with the use of cloud cover during an attack mission. Experiences under these conditions were considered very valuable.

Later during the month, 90th AS spent the second half of October in the field, participating in cooperative missions with ground forces. Nine A-17A's left on 15 October for Fort Leavenworth. The following day 10 more machines joined them at Fort Riley and the following GHQ Air Force demonstrations on the 20th.

A single 90th AS machine on display at the 1938 National Air Races, with a P-35 barely visible in the background. (coll. G. Balzer)

✪ 34th AS lost an A-17A and its crew on 4 January 1938. 2nd Lt. Charles A. Clancy and his passenger, private Victor L. Jost were killed when their aircraft, 36-168, struck power lines while practising formation flying. According to the telegraph report on the accident the weather was excellent and there was no failure of material or equipment. 36-168 was written off on 21 April 1938.

✪ Less than a week later, on 10 January, tragedy struck 90th AS, when A-17A 36-197 was lost during a night navigation flight. The plane crashed at 6:15 pm near Grandview, Tex. 2nd Lt. Frank K. Thompson and Corporal Walter T. Mathews were killed in the accident.

✪ 90th AS lost another A-17A a short while later. On 22 January 36-203 was lost in an accident at Chapman Field, Florida. Pvt. John K. Gahr was commended for his bravery during the aftermath of the accident. Gahr was passenger of the plane and successfully extracted the unconscious pilot from the smoking wreckage.

✪ 35-119, from 1ABS, crashed near Warrenton, GA, on 6 February 1938 and was written off. The pilot, William C. Goldsborough was killed after bailing out.

✪ On 5 July two A-17s from 63rd SchS were involved in a landing accident at Kelly Field. 35-73 was damaged so heavily, that it was struck of charge. 35-127 could be repaired.

✪ 36-259, an A-17A based at March Field, was lost following a mid-air collision on 8 August 1938. The crew could bail out successfully, the fate of the other aircraft is unknown. 36-259 crashed two miles west of Verdemont, California and was written off on 19 October 1938.

✪ A-17A 36-172 was written off on 2 September 1938 after an accident. No details are known.

✪ Chief of the Air Corps, major general Oscar Westover was killed on 21 September, when his personal plane, 36-349 entered a high speed stall after taking off from the Lockheed company airfield at Burbank, California. His passenger, Sgt Samuel Hymes, also died in the crash.

1939 – A YEAR OF STEADY DECLINE

A proposal for a new twin engined light attack bomber had been issued to aircraft companies late 1937, with a dead line to present designs on 17 March 1939. The specifications included a range of 1.200 miles, 200 mph speed, carrying a 1.200 pound bomb load. Contracts were awarded to Douglas, Glenn Martin and Stearman to build a prototype. This new type of aircraft was thought to offer greater survivability over the battle field and could deliver an accurate strike well behind enemy lines if needed. More importantly, 1939 proved to be the final year of first line use for the A-17. The Attack Groups were redesignated as Bombardment Groups and started to fly a mix of B-18 and A-17 types.

A 4 ABS machine.

(coll. G. Balzer)

A large 73rd AS formation. AQ-27 carries an vertical band on the fuselage, indicating it is the commanders' machine.

(coll. E. Hoogschagen)

The red fuselage band indicates that AQ27 was the regular aircraft of the squadron commander. (coll. G. Balzer)

Centre: *36-200, with 11 ABS markings on the tail. (coll. G. Balzer)*

34th AS machines, with the Thunderbird badge painted on the fuselage, lined up at March Field. (coll. G. Balzer)

The 1939 National Defence Day, on 19 March, was held at March Field, home of 17 AG. A large public enjoyed a day filled with demonstrations. Among the activities was a formation of 18 A-17s demonstrating a simulated attack on anti-aircraft artillery.

On 2 August, 13th AS completed a long distance flight visiting 23 towns as part of the Air Corps' 30th Anniversary. Festivities were also held at Chanute Field, which attracted nearly 14000 visitors. As part of demonstrations, a flight of three A-17s made a trip around the field. 95th AS made a more serious 300 mile attack navigation flight, completely flown at tree top level. Part of the exercise was to make use of the landscape as good as possible, taking cover behind hills and flying through ravines.

Tests

The NACA institute used A-17A 36-184 for experiments with engine cowlings. Goal was to find new cowlings for radial engines, providing low drag and efficient cooling. A first attempt resulted in a fully enclosed engine, while cooling was arranged by two air intakes which were built into the leading edges of the wing. This arrangement was no success. The engine overheated during ground runs and the aircraft was never flown with this experimental cowl.

The second attempt utilized a close fitting large propeller spinner which featured a fan system generated enough airflow to cool the engine. This configuration was successful, but did not offer significant improvement over the standard cowling. After experiments were finished, the A-17A was returned to standard configuration and returned to the Air Corps on 21 June 1940.

Above: 36-184 prior to the engine cowl experiments. The photo is dated 22 August 1939.

Centre: The first experimental engine cowl, as photographed on 3 April 1940.

Right: The second variant of the fully enclosed engine cowl. This photo was made on 9 June 1940.

More mishaps		
4 May 1939	*38-353*	Written off after crash at Anacosta, Washington DC.
21 May 1939	*36-245*	95th AS machine, crashed into a observatory building on top of Mt. Hamilton. The crew, 2nd lt. Richard P. Lorenz and pvt. W.E. Scott were killed instantly
23 June 1939	*35-78*	Written off at March Field. Was involved in a landing accident at Kelly Field on 31 August 1938
12 August 1939	*36-192*	Crew bailed out after aircraft suffered structural failure, 4 miles west of Lee Hall, Virginia. The plane was attached to 33rd PS, of 8th PG
20 September 1939	*36-229*	21 ABS plane. Pilot 2nd Lt. J.O. Reed had to bail out after engine trouble. The machine crashed into the McKenzie river near Nimrod, Oregon.
8 October 1939	*35-64*	Destroyed in mid-air collision over Davis Mesa, New Mexico.

An A-17A from 73rd AS escorts a B-17B. A small series of this variant entered service between October 1939 and March 1940.

(coll. G. Balzer)

Centre:
An 17th AG machine visiting McChord Field. Tacoma Field was renamed in honor of Col. McChord, who died in a crash when he tried to perform an emergency landing in his A-17.

(coll. G. Balzer)

NEW ROLES FOR THE ATTACK GROUPS

The 3rd AG was redesignated as the 3rd Bombardment Group (Light) on 15 September 1939. 8th AS traded their A-17s and A-18s for B-18s. Most of the A-17As in inventory were transferred to other, second line, units. 13th and 90th BS retained a small number of aircraft. *36-216* of 13th BS was lost after a mid-air collision on 10 February 1940, killing its crew.

Similarly, 17th AG received the bombing role and was redesignated to Bombardment Group (Medium) on 17 October 1939 and received B-18 bombers. A handful of A-17s and A-19s remained with the Group. *35-128* and *35-129* continued to fly with HQSQ of 17 BG. The A-19s were used as target tug with 95th BS. The A-17s were respectively struck of charge on 10 June and 8 February 1942.

Many A-17s were transferred to other units, a lot of which were operating in the Panama Canal Zone. Others were scattered around all sorts of units in the United States itself. During the early months of 1940 fifteen A-17s and A-17As were loaned from Hamilton, March and Langley Fields to the flying school at Randolph Field, where BT-13 and BT-14 trainers were expected. The Northrops were a welcome stop gap solution to the fleet of trainers. Aircraft were reassigned frequently to other units, making it nearly impossible to reconstruct a (comprehensible) chronological time line. For this reason, most of this chapter proceeds by describing individual units.

Secretive transfer of aircraft

Attack planes from the 3rd AG swoop down on a truck column. Instead of lethal mustard gas a thin solution of white wash is being sprayed. The results of the chemical attack could be determined by the white specks found on trucks and soldiers. (Coll. E. Hoogschagen)

On 6 June 1940 orders were issued to transfer 92 A-17As to Langley Field. This order reached no less than 19 airfields and had come directly from the offices of the Chief of the Air Corps.

Edgewood Arsenal	1
Aberdeen	1
Langley Field	11
Randolph Field	1
Kelly Field	1
Brooke Field	1
McDill Field	1
Mitchell Field	1
West Point	1
Chanute Field	1
Fort Leavenworth	1
Maxwell Field	2
Scott Field	3
Barksdale Field	9
Lowry Field	3
March Field	16
Moffett Field	5
Wright Field	13
Bolling Field	2

The orders should be completed before 11 June and specifically mentioned that no publicity be given to the transfer. Guns and gunsights had to be removed, but all other armament (i.e. bomb racks) could remain. On 7 June an alternative order had reached the Air Fields involved. The aircraft should be distributed between Langley Field, Mitchel Field and Middletown Air Depot. This was to reduce publicity as far as possible. Also, all armament should remain fitted in the aircraft. 13 aircraft at Langley Field were now to be transferred to Middletown Air Depot. The aircraft at Edgewood, Aberdeen, West Point and Bolling Field were to transfer to Mitchel Field. During the transfer flights, all nationality and unit markings were removed. Only a small civil registration was applied on the tail rudder. Spare parts were to be concentrated at Middletown Air Depot.

This transport would be arranged by 10th Transport Group. The hush-hush transfer of aircraft was part of secret aircraft deliveries to France, which needed combat aircraft desperately. (see next chapter)

After the transfer was completed, 95 A-17s and 24 A-17As remained in service. These were, according to a stock survey of 30 June 1940, spread over no fewer than 32 airfields and depots.

Left:
The O-3 markings may have been temporarily applied.
(coll. G. Balzer)

35-122 as flying test bed

The *35-122* was sent to NACA at Langley and was used as flying test bed for tests with laminar flow wings. At first the right wing was fitted with a 1,5 m wide new wing section. In a later phase of testing, both wings received highly polished wide chord wing sections. Small auxiliary engines were placed in front of the leading edge to generate air flow. After disappointing results, further testing of laminar flow wings was undertaken in wind tunnels. *35-122* was returned to original configuration and returned to the Air Corps. (coll. G. Balzer)

The first produced A-17A lasted well into at least October 1940, when the Air Corps started to paint their aircraft in olive drab (no 41) and neutral grey (no 42). (coll. G. Balzer)

The checkerboards were still an extravagant appearance during the late thirties, but would be a very common sight on US combat aircraft during the second world war. Profile by Alexandre Guedes)

Staff Squadron aircraft were painted with a blue and yellow checker board. (coll. G. Balzer)

35-117, attached to Staff Squadron, hit a car during take-off on 25 May 1940. (Coll. E. Hoogschagen)

.... and more mishaps		
10 February 1940	36-216	13 BS. Lost after mid-air collision, crew killed.
05 March 1940	35-71	11 ABS machine. Written off after landing accident at Randolph Field
10 March 1940	36-183	Crashed 4,5 miles northeast of Sharon Springs, Kansas. Pilot Major Devereux M. Myers and Corperal Maurice E. Melvin were killed. The plane was attached to 21 ABS from Albequerque, New Mexico.
22 March 1940	36-163	Written off in accident
07 April 1940	36-173	9 ABS. Crew bailed out after running out of gas near Tyler, Texas.
22 April 1940	36-190	Written off after accident (which may have taken place earlier, on 29 February)
23 April 1940	35-87	Written off after take-off accident at Mitchel Field
01 May 1940	36-169	Written off after stalling near Moscow, Idaho. Crew Killed.
25 May 1940	35-117	3rd Staff Squadron machine. Crashed during take-off from Dodge City, Kansas, killing the crew.
6 August 1940	35-142	Crashed at France Field, crew killed.
21 August 1940	35-107	Written off following landing accident at Camp Skeel, Michigan.
28 September 1940	35-137	Struck off charge
8 October 1940	35-160	Written off after crash near Lake Point, Utah
19 October 1940	35-109	Crashed near Alta, Utah. Crew killed.

A large number of AB squadrons operated A-17s at any given time. This started in 1936 and lasted until the final machines were phased out.

1 ABS   Langley Field
2 ABS   Mitchel Field
3 ABS   Selfridge Field
4 ABS   March Field
5 ABS   Hamilton Field
6 ABS   Barksdale Field

35-150 while serving 1st ABS. Little is known about its operational use.

(coll. G. Balzer)

8 ABS   Scott Field
9 ABS   Moffett Field
10 ABS   Chanute Field
11 ABS   Randolph Field
12 ABS   Randolph Field
16 ABS   France Field
20 ABS   Pendleton
21 ABS   Lowry Field
30 ABS   Morris Field / Charlotte AAB
32 ABS   Tucson army air field
39 ABS   Key Field
42 ABS   Dale Mabry Field
44 ABS   Portland AAB
45 ABS
46 ABS   Hamilton Field
54 ABS   Lawson Field
59 ABS   Keesler Field
336 ABS Rio Hato
432 ABS Bedford AAB

2nd ABS aircraft stationed at Mitchel Field.

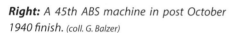

Right: *A 45th ABS machine in post October 1940 finish.* (coll. G. Balzer)

The 7th BG received a number of A-17s during 1939. This unit was based at Hamilton Field and later moved to Fort Douglas, Utah. Each of the squadrons used a machine for pilot proficiency. When the A-17As were phased out from the first line squadrons some found their way to the 7th BG as well. Very little is known about A-17 operations. A single (unknown) A-17 is reported as part of inventory of 88th Reconnaissance Squadron (RS).

An A-17 carrying tail code BB, indicating that it was attached to the 2nd BG. (coll. G. Balzer)

This machine carries the kicking mule badge of 88 RS. The cowl front is painted in red and yellow. (coll. G. Balzer)

A 5th ABS A-17 with the squadron badge on the fuselage; a blue disk with 5 aircraft silhouettes, parked tail to tail.
(coll. G. Balzer)

Right: Despite being an aircraft used by 6th ABS, this machine is fitted with gas tanks and visor on the nose.
(coll. G. Balzer)

35-77 when it was attached to 32nd ABS. (coll. G. Balzer)

35-106, flying with 11 BS, was involved in a mid-air collision during formation flying on 16 October 1939. The other plane involved was 35-160. Both aircraft, although damaged, were able to return to their airfield. 35-160 crashed after mechanical failure on 8 October 1940, when it still served with 11 BS. It was subsequently written off. 22nd BS was formed on 20 October 1939 and among the first planes were two A-17s and a single A-17A. The Northrops were used as squadron hacks.

Hamilton Field hosted an open house on 21 January 1940. 88th RS exhibited Aerial and Ground photography equipment and among the aircraft on display were an A-17A and a big Boeing B-17B four engine bomber. On 23 September 1940 an unknown A-17 was involved in a forced landing after it suffered engine failure while flying over salt flats in Utah. 35-106 was finally written off on 9 March 1941 when it flew with 46 ABS. The pilot, Roy P. McDonald had attempted a wheels down landing on a beach at Coos Bay. He was killed in the accident, the observer survived. 35-109 was lost on 7 October 1940 when it flew into ground at Alta, Utah. The crash claimed the life of its crew. This plane served with 22nd BS. 35-123 flew with 9th BS and ground looped at Salt Lake City airport on 8 December 1940.

35-127 crashed in a similar way on Fort Douglas airfield on 20 February 1941. 35-61, attached to 22 BS was lost on 4 April 1941 after a mechanical failure. The crew successfully bailed out near Devils Slide, Utah.

Two A-17s were part of the 29 BG manifest, which was the first unit to use the newly constructed MacDill Field, near Tampa, Florida. The unit arrived on 15 May 1940, and A-17s were used until at least 27 March 1942. On this day 35-155 was destroyed when the plane crashed into the Gulf of Mexico, killing its crew 2nd lt. Thad P. Medlin and 2nd lt. Joseph E. Anderson.

At least ten A-17s and A-17As were used by 19th BG, prior to their departure to the Philippines. 19 BG was stationed at March Field, and moved to Albuquerque, New Mexico during April 1941. Aircraft served pretty uneventful with 30th, 32nd, 93rd

Interestingly, this A-17 carries tail-code BG from 7th BG, while the squadron badge is from 9th BS.

Below:
An A-17 with an alternatively styled plane number, possibly while attached to a bombardment squadron.

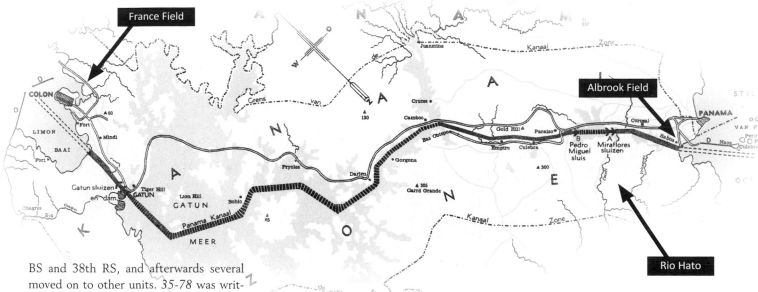

BS and 38th RS, and afterwards several moved on to other units. *35-78* was written off after ground looping during landing on 23 June 1939. *35-72*, flying with 32nd BS was involved in a ground collision on 3 February at March Field and had to be written off. A few days later, February 7th, *35-124*, attached to 30th BS was written off after a landing accident at Palm Springs. *35-146* was involved in a collision with a bird on 20 October 1941. It would be struck off charge on 1 January 1942. *35-112* was lost on 1 November 1941. It was attached to 38th RS which was based at Albuquerque, New Mexico. It crashed in Bear Canyon, 20 miles north east of its base. The crew was killed in the impact.

Another two A-17s were used by 6th RS, part of 41st BG, when this Group was operating from Mather Field, Sacramento. One, *35-94*, had earlier served with 93rd BS and was finally struck of charge on 26 March 1943. 6th RS was redesignated as 396th BS on 22 April 1942 and was tasked to patrol the Californian Pacific coast. The second machine, A-17A *36-186*, had previously served with 90th AS and is last mentioned on 25 June 1943, when it crashed near Marshall Field after an engine fire.

PANAMA CANAL ZONE

Sending A-17s to the Canal Zone had been contemplated for a long time. This finally became possible when the improved A-17A started to reach the units stationed in mainland USA. The first Canal Zone unit to receive A-17s was 74th AS, which was part of 16th PG. The unit had switched from the pursuit role and was redesignated 74th AS on 1 September 1937 but kept flying their Boeing P-12s for a little longer.

Home base was Albrook Field. Shortly afterwards, 14 A-17s reached Albrook Field. The men were quickly called upon, when an aerial search for a missing fishing boat was conducted on 28 September. Eight A-17s, divided in four patrols of two airplanes each, patrolled the coast line to ten miles off coast. The fishing boat was found, adrift, after a three hour search period.

A 74th AS machine peels off. This machine seems to be unarmed, except for the bomb racks.
(coll. G. Balzer)

74th AS men constructed the necessary target structures at Rio Hato which was completed by the end of October. Ten wooden targets were constructed, fifty feet long and three feet high. The completed gunnery range was taken into service during the following month.

The 74th AS completed their 1938 season on 27 September with a successful demonstration of an attack on an anti-aircraft battery. A single machine screened the battery in smoke, while following elements of three aircraft attacked with machine gun fire, bombs and chemicals.

74th AS started 1939 with a three day exercise, cooperating with US Navy ships which were planned to pass through the Panama Canal. Together with other units based at Albrook Field and France Field, 74th AS provided patrol duties over the fleet on 13 and 14 January. Main goal was to test communication equipment and procedures during a patrol task, including the cooperation with other (naval) units.

Albrook Field was undergoing extensive renovations during this period and received a new main and secondary runways, hangars and aprons. Flying from the new runway commenced on 17 April 1939.

74th AS was redesignated as Bombardment Squadron on 1 November 1939. It kept flying their A-17s for a while longer. During 1939 two A-17s were transferred to 16th PG Group Headquarters squadron. On 21 November '39 74th BS sent nine A-17s and a single B-18 to Tegucigalpa, Honduras, while transporting members of the Congressional party. A single A-17 had picked up a VIP at San Jose, Puerto Rico. The formation of aircraft set out for Managua, Nicaragua, where fuel was taken in. The aviators were welcomed with festivities and a banquet at Honduras City. The

A formation of 74th AS planes, led by a differently marked machine. Perhaps this was an ABS plane. (coll. G. Balzer)

Flying the colors

During February 1938 the 19th Wing, Panama Canal Department, left Albrook Field and France Field for a massed flight to Guatemala city for an official visit to this country. The formation was commanded by Brigadier General George H. Brett, and counted various aircraft types. The command aircraft, an Sikorsky Y10A-8, flew non-stop to Guatemala City. A-17s and P-12s left Albrook Field during the morning of 7 February and landed at Managua, Nicaragua for an overnight stop. During the evening, the air corps pilots attended a reception given at the president's palace, followed by a reception given by the American ambassador. While the single seat aircraft left Managua the following morning, Martin B-10s departed from France Field for a non-stop flight to Guatemala. The entire formation made a rendez vous at Escuintla, and proceed to Guatemala City.

The crew were formally welcomed at the field, after which a day long programme of reviews, demonstrations, meetings followed. An evening party concluded the day. The following day the lined up aircraft were reviewed and an aerial review, in honor of Guatemalan president Jorge Ubigo, passed over the Campo del Marte reviewing grounds. The day ended with an evening banquet. Early Friday mor-

ning, the entire flight departed from Guatemala City. The twin engine aircraft flew non-stop to their home stations, while the single engined machines flew to San Jose (Costa Rica) and the crews spent the night there. All aircraft returned to Albrook Field on 12 February. The flight and visit to Guatemala was an outright success.

Servicing aircraft in blistering heat was very common, but the men frequently had to deal with tropical storms and flooding as well. (coll. G. Balzer)

Tail number 51 is being pushed back in a hangar. The plane carries the 74th AS' badge, a frigate bird on a blue field.

(coll. G. Balzer)

Below: *An A-17 belonging to 6th BG. The plane carries the 74th AS badge and a red cowl front.*

(coll. G. Balzer)

flyers stayed the night at the Army School of Aviation of Honduras and all safely returned to Albrook Field the following day. On 1 February the squadron was reassigned to the 6th BG.

Other A-17s were attached to units of the 6th BG when it was stationed at the Panama Canal Zone. 7th RS was attached to the 6th BG during October 1939. Besides the Douglas B-18 the unit flew at least one A-17. 25th BS, which had a long service record at France Field, also received at least one A-17. A fatal incident involving *35-82* took place on 9 July 1940, when this aircraft crashed into the Caribbean Sea. 3rd BS, flying with a mix of B-17Es and B-18s at Rio Hato also received a number. The first, *35-130*, was on strength during January 1942 but was written off after a take-off accident at Chame on 27th of the same month. During mid-February another A-17 was added. Two A-17s (*35-132* and *35-138*) were still used as hacks during the 3rd BS deployments to Talara, Peru and Salinas, Ecuador. The aircraft may have been in poor technical shape, as the aircraft were left stranded at Cali, Colombia, when 3rd BS returned to David Army Airfield, Panama. *35-84* Was mentioned in three separate accident reports in the period January up to September 1942 when it was used by 6 BG HQSQ. *35-131* also flew with this unit. It was struck of

charge on 12 November 1942 after a forced landing near Boquete on 24 July. *35-132* served with 336 ABS on Rio Hato. It is last mentioned when it crashed during landing here, on 20 July 1942. The last A-17 serving with 6 BG may have been *35-134*, which was involved in a take-off accident at Albrook Field on 11 September 1943. It served with 397th BS (formerly 7th RS).

39th OS temporary flew a number of A-17s after this unit was activated on 1 February 1940. Six A-17s were flown while waiting for its assigned North American O-47A observation planes. A detachment of six crews from 39th OS joined a large

formation flight to Guatemala City. The men were given the opportunity to fly A-17 planes for this trip. The formation left France Field on 28 June for a refuelling stop at Managua, Nicaragua. The next day the men proceeded to Guatemala City where all aircraft involved displayed in a mass formation. Festivities followed, honouring Guatemalan independence. The return trip was made on 30 June, via San Salvador for a nights rest. All six aircraft arrived safely at France Field the following day, after a tiresome 6,5 hour non-stop flight. 39th OS later became assigned to 72nd OG (Observation Group) and a number of A-17s were kept in inventory up to January 1942.

At least two machines were flying with the 29th BS, 40th BG. This group was previously stationed at Borinquen Field, Puerto Rico. On 22 October 1942 *35-85* crashed during take-off at Aguadulce due to mechanical failure. *35-138* was last mentioned when it ground looped after landing on 6 November 1942.

Several staff units received A-17s. During early 1940 at least two were stationed at France Field, serving with HQ squadron and 16 ABS. *35-142*, while serving with the latter, was destroyed in a fatal crash on 6 August 1940. *35-89* Was written off after a landing accident on 21 June 1941. Similar numbers served with from Albrook

Field, although little is known. *35-130* flew with HQSQ of 18th PG. It was involved in a landing accident on 3 December 1940 but was repaired and transferred to 3rd BS. *35-90* was used as hack with 53rd PS, 32nd PG. It went missing on 1 November 1941. Its crew, pilot David Chaimovich, was reported missing in action. It is not known if there was an observer aboard. *35-86* served with HQ squadron at Albrook Field and was finally written off after a landing accident at Albrook Field on 31 July 1943. 37th PG was activated at Albrook Field on 1 February 1941. Besides 25 Boeing P-26As, the unit received two BC-1 trainers and a similar number of A-17 aircraft.

18th RS, part of 9th BG, received at least two A-17s when it was initially based at Mitchell Field. *35-145* and *35-155* were temporarily used here. On 20 November 1940 the Group moved to Rio Hato, Panama. Here, another A-17 is mentioned. *35-136* was first used by HQSQ of 9th BG, and was transferred to 59th BS, 25th BG, as target tug. On 2 January 1942 the machine was lost in a crash in Parita Bay, Panama.[1] 44th RS, which was based at Howard Field operated at least a single A-17 during 1941.

---

1 Although 59th BS was very actively involved in the hunt for German U-boats, no sources indicate that A-17s were involved in actual combat duties.

Two A-17s at Borinquen airfield.
(coll. E. Hoogschagen)

Puerto Rico

Borinquen Field, located on Puerto Rico, was a second overseas station which hosted a unit operating A-17s. Construction of this field started during the first week of September 1939 and the first Air Corps detachment settled down 21 November. The first plane, a B-18 from 27th RS, landed on 27 November. An A-17A was attached to the squadron. This machine, *38-348*, was involved in a landing accident on 28 April 1940. The 25th BG, which was activated on 1 February 1940 at Langley

Field, would be permanently stationed at Borinquen Field. 10th, 12th and 35th BS were attached to the Group, and 27th RS was assigned to the Group soon afterwards. The first aircraft, from 35th BS, arrived 3 November 1940. This unit operated 14 B-18s and two A-17s. 33 and 34 Material Squadron (MS) also joined the Group.[1] On 29 January 1941, not long after arri-

---

1 During April 1941, 25th BG was split in half to form 40th BG, with 29th, 44th, 45th BS and 5th RS

val, *38-355* crashed into the sea, south of Punto Soldado. This machine was attached to 33 MS. One of the aircraft attached to 35th BS was *35-56*, which was involved in a take-off accident on 18 July 1941 and subsequently written off. *38-348*, which had been transferred from 27th RS to 10th BS was lost on 26 November 1941 after a mid-air collision. *35-95*, also serving with 10th BS was written off after a landing accident at Borinquen Field on 28 February 1942.

THE A-33

The 8A-5N variant was the result of requirements issued by the Norwegian government to equip its Heerens Flyvevaben (Army Flying Service) with an A-17A powered by the 1,200hp Wright R-1820-G205 "Cyclone" engine. The resulting variant, identified as 8A-5N, was outwardly identical to the previous 8A-3P, delivered to Peru, and 8A-4, delivered to Iraq, models but the more powerful engine allowed heavier armament to be carried. A pair of Browning .50 caliber machine guns were placed in aerodynamic fairings below the outer wings, raising the number of fixed weapons to six while the bomb load reached 820kg. 36 airframes were ordered in early 1940 but none reached Norwegian soil as they were embargoed by the US government when Norway fell to the Germans in June 1940. When Norway established a government in exile in the United Kingdom at the end of 1940 and its reformed air force established a training center on Island airport, in Toronto,

A crewman carries a belt of .50 ammunition to an A-33. The planes did receive fiscal numbers but also a temporary Z code on the fuselage.

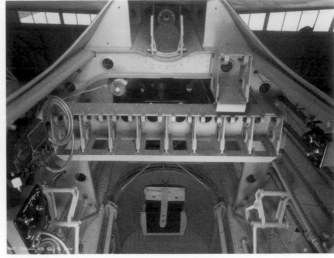

Many of the Douglas 8A variants which were exported featured a retractable observers gondola, which gave more work space for the observer when a bomb run was being made. At left the gondola is seen in retracted, and at right in lowered position. The bomb sight could be fitted in the gondola.

Right:
Two photos showing the .50 Browning machinegun in the underwing gun pod. The ammunition box was placed below the gun

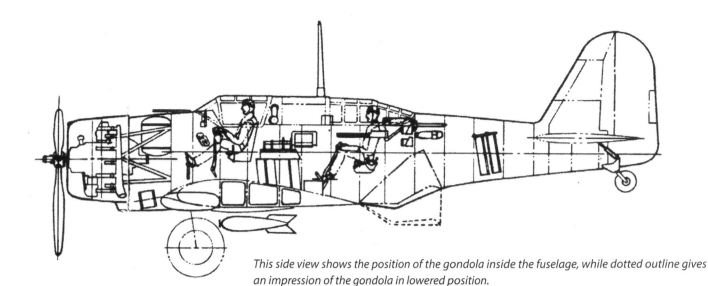

This side view shows the position of the gondola inside the fuselage, while dotted outline gives an impression of the gondola in lowered position.

Canada. The Douglas and other aircraft purchased by the Norwegian government were sent there to serve as trainers. At "Little Norway", as this base was known, these aircraft served until mid 1941 when 18 surviving machines were repossessed by the US government on a reverse lend lease, and pressed into USAAF service as Douglas A-33-DE.

As the USAAF found no use for these aircraft the US government considered their sale to the URSS at a time, but this came to nothing and the surplus airframes were sent to the USAAF storage facilities at Kelly Field, in San Antonio, Texas.

The aircraft were put in US Air Corps service two days after the attack on Pearl Harbor and were serialled *42-13584* up to *42-13601*. The aircraft were known as A-33.

The aircraft were mainly operated by the 352nd Flexible Gunnery Training Squadron at Las Vegas and 50th, 51st, and 476th School Squadrons. The first two were located at Las Vegas, the latter at Harlingen. *42-13590* was struck of charge after a landing accident on 1 May 1942 at the Tonopah Gunnery School. Three days later, *42-13591* was a total loss after a fatal crash 10 miles north west of Las Vegas Air Gunnery School. *42-13593* was lost in an accident on 3 May 1943.

Technical data A-33-DE (8A-5N)	
Engine	Wright GR-1820-G205A Cyclone
Power	1200 hp
Wingspan	47 ft 9 in (14,55 m)
Length	32 ft 6 in (9,91 m)
Height	12 ft (3,66 m)
Empty weight	5,510 lb (2.499 kg)
All-up weight	8,600 lb (4.173 kg)
Max. speed	248 m/ph (399 km/h) at 15,700 ft (4.800 m)
Range	910 m (1.464 km)
Service ceiling	29,000 ft (8.800 m)
Climb rate	-
Armament	4 fixed 0.30 in wings, 2 0.50 in in underwing pods, one flexible 0.30 mg
Bombload	20 internally stored 30 lb (13,6 kg bombs plus four externally carried 100 lb (45 kg) bombs or flares. or four 500 lb (230 kg) bombs externally or eight 100 lb (45 kg) bombs externally

This photo from the manual shows a bomb load of eight 100 lb bombs.

A Douglas DB-8A5N after it was taken into US service as A-33.
(Coll. E. Hoogschagen)

The fiscal number on the tail of 42-13594

An added loop antenna is clearly visible on this old magazine photo. This antenna was often added to Staff squadron machines.

Three other machines were struck of charge or condemned before the bulk of the aircraft were taken out of service on 26 September 1944. The very last A-33s were withdrawn from use on 16 February and 31 March 1945.

In June 1943 Thirteen surviving Norwegian machines were finally transferred to the Air Corps and were serialled *42-109007* to *42-109019*. Contrary to the earlier aircraft intended for Peru, these machines were delivered to Peru between July and November 1943.

Mishaps 1941		
29 January 1941	*38-355*	Crashed into the sea, south of Punto Soldado, Puerto Rico
3 February 1941	*35-72*	Ground collision at March Field, crew OK. 32nd BS, 19th BG.
7 February 1941	*35-124*	Written off at Palm Springs, California, after a landing accident. 30th BS, 19th BG
21 February 1941	*36-261*	Crashed near Athens, Ohio, after mid-air collision. Crew killed
9 March 1941	*35-106*	Crashed during forced landing on beach at Coos Bay, Oregon. Pilot killed, observer survived. 46th ABS
20 March 1941	*36-206*	Written off after crash
4 April 1941	*35-61*	Crew had to bail out near Devils Slide, Utah. Struck off charge 29 April. 22nd BS, 7th BG.
12 June 1941	*35-110*	Written off in fatal forced landing accident, Selfridge Field. 20th ABS
8 July 1941	*36-241*	Fatal crash near Hayees Store, Virginia. The plane was attached to 2nd Staff Squadron
27 June 1941	*35-66*	Written off after forced landing at Truckee airport. 44th ABS
25 July 1941	*35 65*	Written off after crash at Goldendale, Washington State.
7 September 1941	*35-149*	Accident at Greencaste, Indiana. Crew killed. Written off 29 September 1941. 8th ABS
17 October 1941	*35-89*	Written off after anding accident, France Field
25 October 1941	*35-76*	Written off after crash at Morgantan, Georgia. 59th ABS
5 November 1941	*35-90*	Missing after flight from Albrook Field. Crew reported missing in action. 53rd PS, 32nd PG
17 November 1941	*35-112*	Accident at Bear Canyon, written off 4 February 1942. 38th RS, 19th BG
26 November 1941	*38-348*	Heavily damaged in mid-air collision, Aquadilla, Puerto Rico. Subsequently written off on 31 January 1942.
13 December 1941	*35-147*	Written off, crashed near Richmond, Indiana. Crew was killed attempting to bail out.
26 December 1941	*35-70*	Written off after crash in Conococheague Moutains, near Blain, Pennsylvania. Crew killed.

Left: *A parachute is being attached to an A-17s bomb rack for a drop test at Maxwell Field, 23 October 1941.*

35-157, seen here at Hammer Field.

Mishaps 1942		
2 January 1942	35-136	59th BS. Crashed in Parita Bay, Panama, crew Leroy W. Smith killed.
7 January 1942	35-81	Ground collision at Wilmington, North Carolina. The plane was written off on 5 May 1942.
27 January 1942	35-130	3rd BS. Take off accident after engine failure at Chame, Panama.
28 February 1942	35-95	10th BS. Written off after landing accident at Borinquen Field, Puerto Rico.
10 March 1942	35-151	Attached to 42 ABS. Written off following landing accident at Cochran Field on 30 January '42.
27 March 1942	35-155	29th BG. Crashed into the Gulf of Mexico, crew 2nd lt. Thad P. Medlin and 2nd lt. Joseph E. Anderson killed.
2 May 1942	35-145	39 ABS. Written off following landing accident at Lambert Field, Missouri.
17 July 1942	36-231	12 ABS. Written off following landing accident at Randolph Field, Texas.
24 July 1942	35-131	6 BG. Crashed 2 miles west of Boquete, Canal Zone. Written off on 12 November 1942.
10 October 1942	35-69	Written off after landing accident at Towanda airport.
22 October 1942	35-139	54 ABS. Involved in fatal landing accident at Lawson Field, Georgia.

36-257 with SAD marking, indicating it was stationed at the San Antonio Depot.

35-58, as it appeared before 6 May 1942, when the red dot was omitted.

35-77, seen with the roundel which was carried between 6 May 1942 and 28 June 1943. It was involved in an accident on 15 July 1942, when it was hit by another machine which was taking off.

35-79 served until after 6 May 1942. Its eventual fate is not known.

Mishaps 1943		
29 January 1943	35-067	Crew had to bail out after the aircraft entered a spin, 10 miles west of Yakima, Washington. May have been attached to 17th BG.
?? April 1943	35-91	Written off. Details unknown.
22 April 1943	35-108	Ground looped after landing at Hunter Field, Georgia.
19 May 1943	36-240	Involved in a take-off accident at Edgewood Arsenal. Written off on 25 September 1943.
14 July 1943	35-157	Written off after take-off accident at Camp Davis, North Carolina.
28 July 1943	35-58	Written off after landing accident, Sherman Field, Kansas.
31 July 1943	35-86	Written off after landing accident at Albrook Field, Panama.
11 August 1943	35-118	Ground looped after landing and subsequently written off at Raco airport.
13 August 1943	36-257	Destroyed by fire after landing accident at McLellan Field.
11 September 1943	35-134	397 BS. Written off after take-off accident at Albrook Field.
06 January 1944	36-164	Crashed after entering a spin at Aberdeen Proving Ground, crew killed.
17 May 1944	35-115	Written off after forced landing 4 miles east of Mount Horeb, Wisconsin.
17 May 1944	35-121	As above - landed 2 miles east of Mount Horeb.
30 June 1944	35-060	Written off after forced landing due mechanical failure, Miami Florida.
27 July 1944	36-234	Written off after engine fire, Moorefield, West Virginia

Records seem to be contradictory, since (at least) 19 out of 110 A-17s were written off prior to 30 June, while the fate of other aircraft involved in severe crashes is still unclear. (At least) 16 out of 129 delivered A-17As were lost as well. The exact figures may differ from the inventory presented in this text. It may be possible that new aircraft were constructed out of spares.

Little is known about the service life of 35-63, but it is seen here with the white bars added to the roundel. The whole was outlined in red. This was introduced on 28 June 1943.

Above: *This A-17A carries the Capitol Dome insignia of the 14th BG*

A staff aircraft of GHQAF (General Headquarters Air Force) carries the winged star on the fuselage

Failed export to France

By Edwin Hoogschagen

After the German invasion of 10 May 1940 French prime minister Reynaud, whom had been installed in office barely two months earlier, pleaded the United States for the immediate assistance to France. President Roosevelt sent out instructions to release aircraft for purchase by France. Among these were 93 USAAF Northrop A-17As. Purchase of the fixed undercarriage A-17 was discussed, but this did not materialize. Contract no. 143 was dated 15 June 1940. France layed down its arms the following day, and simultaneously, the French Purchasing Commission transferred all outstanding contracts to the British Purchasing Commission. Contract no. 143 now became contract F-691.

A symbolic photo, the USAAC tail flash is being overpainted.

(Coll. E. Hoogschagen)

Above and middle:
Aircraft amassed with temporary civil registrations.
(Coll. G. Balzer)

On paper, the aircraft were actually bought by Douglas and resold to France to avoid breaking the US neutrality laws, which strictly forbid the delivery of USAAC inventory to foreign powers. The aircraft were flown to the east coast with temporary civil registrations *NX-N1* to *NX-N93*. The machines were flown to Middletown Air Depot and Mitchel Field. From there, the planes continued to Halifax, Nova Scotia, the first of them arriving there on June 21.

Plans had been made to have the aircraft transported to France by aircraft carrier *Béarn*, which had arrived at Halifax on June 1. It had carried 164 tons of gold bullion to safety. On its return trip, It would carry a load of hastily acquired combat aircraft back to France.

When the aircraft arrived in Halifax, it turned out that *Béarn* had already departed. She had left a few days earlier, on June 16th, together with her escort, the light cruisers *Jeanne d'Arc* and *Emile Bertin*. During the past days, the defence of France had collapsed, and the vessels had set course for North Africa and eventually diverted to the island of Martinique where they arrived on June 27.

GREAT BRITAIN

After taking over the French order the aircraft received Air Ministry serials *AB541* to *AB633*, but these were never actually applied. The aircraft were loaded on British merchant ships and the first machines reached Great Britain during July. Serials were allotted as follows:

This page:
AS441 photographed at Boscombe Down.
(SAAF Museum Collection - Via K. Smy)

Serials RAF	
AS440 – AS462	Unloaded at Liverpool, date of arrival unknown
AS958 – AS976	Unloaded at Glasgow, 10 July 1940
AW420 – AW438	Unloaded at Glasgow, 25 July 1940

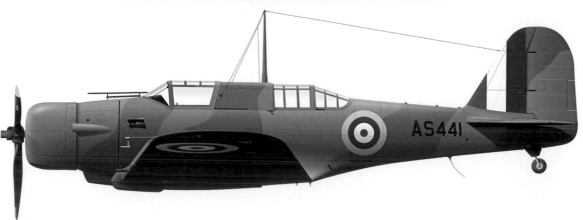

AS441 was test flown at Boscombe Down, after which it was decided that the type was not usable for front line use. No formal report was written when the tests were concluded. The engine in particular was regarded unsuitable for war duties and the planes never reached RAF service. By that time 61 out of 93 aircraft had been shipped. The two latter batches were assembled by Scottish Aviation Ltd (SAL). A surviving SAL job sheet excludes *AS974* to *AS976*, yet all reached South Africa safely. Test flight of the first assembled machines was reported on 1 August and afterwards the aircraft were sent to various maintenance units; 9 MU at Cosford, 20 MU at Aston Down and 48 MU at Hawarden. This whole process was completed on 28 September.

Shipment of the remaining aircraft which were still at Halifax was cancelled. These were made available to the RCAF. Like many aircraft made in the United States, the type was given a nick name: Nomad Mk.1, yet a British Air Commission report dated 15 March 1941 mentions the designation Northrop A-17-A.

All but one of the aircraft were transferred to the South African Air Force. The first Northrop, *AW420*, was shipped on 17 November 1940. The final shipment left for South Africa on 29 July 1941. *AW421* remained in the UK and became Instructional Airframe *2670M*.

AS441 photographed at Boscombe Down. (SAAF Museum Collection - Via K. Smy)

Centre and below:
AS974 was photographed during August 1940.
(coll. P. Butler)

Shipping Date	Serial Numbers (RAF)							Arrived SA	Lost at Sea
17-nov-40	AW420							1	
23-nov-40	AS445							1	
28-nov-40	AS454	AW424						2	
9-dec-40	AS452							1	
16-dec-40	AS450	AS451	AS453	AS455	AS975	AW425			6
17-dec-40	AS447	AS959	AW429					3	
19-dec-40	AS974	AW432	AW438					3	
21-dec-40	AW430	AW435	AW437					3	
24-dec-40	AS969							1	
31-dec-40	AS965							1	
15-jan-41	AS458	AS960	AS963	AS964	AW427	AW428		6	
1-feb-41	AW426							1	
18-feb-41	AS459	AS460	AS961	AS448	AS962	AS968			6
26-feb-41	AS457	AW431	AW433					3	
9-mrt-41	AS972	AW434						2	
14-apr-41	AS461							1	
16-apr-41	AS441							1	
17-apr-41	AS442							1	
30-apr-41	AS440	AS449	AS967	AS971	AW423				5
27-jun-41	AS966	AS976						2	
30-jun-41	AS446							1	
29-jul-41	AS456							1	
NO DATES	AS443	AS444	AS970	AS973	AS462	AS958	AW422	8	
	AW436								
TOTALS								43	17

Shipping details and research by South African researchers K. Smy and W. Brent

✪ SS *Barneveld*, carrying six A-17s, left the port of Hull. She was captured by the pocket battleship *Admiral Scheer* on 20 January. The ship was sunk by explosives the following day.

✪ SS *Grootekerk*, carrying six A-17s, left the port of Swansea on 18 February and was torpedoed by *U123* on 24 February, with all hands lost.

✪ MV *Clan McDougall*, carrying five A-17s, left the port of Glasgow on 30 April and was torpedoed by *U106* on 31 May 1941.

SS Barneveld

SS Grootekerk

MV Clan McDongall

TRANSFER TO SOUTH AFRICA

Of the aircraft which were diverted to the UK, 60 were transported to South Africa. Exact date of arrival is not known, but on 10 March 1941, 17 were being assembled at Port Elizabeth. They were intended for No. 42 Air School. One machine was ready for acceptance and would be used to get acquainted with the type at Maintenance School.

15 April 1942. 1250 hit a tractor whilst taxiing. The crew, Sgt/Pilot RH Durham (RAF), LAC JR Parnham (RAF) escaped without injury. The tractor was operated by 41 Air School. Perhaps the aircraft payed a visit to East London. (SAAF Museum Collection - Via K. Smy)

RAF	SAAF	
AS440	1258	LOST AT SEA
AS441	1263	
AS442	1228	
AS443	1267	
AS444	1229	
AS445	1231	
AS446	1249	
AS447	1240	
AS448	1252	
AS449	1259	LOST AT SEA
AS452	1247	
AS454	1232	
AS456	1248	
AS457	1253	
AS458	1222	
AS461	1264	
AS462	1265	
AS958	1243	
AS959	1238	
AS960	1223	
AS963	1224	
AS964	1225	
AS965	1245	
AS966	1250	
AS967	1260	LOST AT SEA
AS969	1237	
AS970	1268	

RAF	SAAF	
AS971	1261	LOST AT SEA
AS972	1257	
AS973	1269	
AS974	1239	
AS976	1251	
AW420	1227	
AW422	1226	
AW423	1262	LOST AT SEA
AW424	1230	
AW426	1246	
AW427	1221	
AW428	1244	
AW429	1235	
AW430	1234	
AW431	1255	
AW432	1241	
AW433	1254	
AW434	1256	
AW435	1233	
AW436	1266	
AW437	1236	
AW438	1242	

17 Machines were lost at sea while en-route to South Africa [1] – 43 aircraft entered SAAF service. The aircraft received serials *1221 - 1251, 1253 - 1257* and *1263 - 1269*.

Northrop related subjects were introduced in the SAAF training syllabus in the first week of June 1941. During this very same week, two machines were lost in accidents.

The Nomads were to be used as advanced trainer at No. 41 and No. 42 Air Schools. They were used for observation courses, and also for air gunner and air bomber courses. A number of aircraft would be used as target tug. At least 16 aircraft were modified for this purpose and were called Northrop TT. The Northrop could be fitted with a Grumman winch for towing a three meter (10 ft) long drogue. SAAF training command was equipped with British material: Harts, Battles, Masters, Oxfords etc. The Northrop was the only American made aircraft in inventory. This would be the beginning of a somewhat troublesome career in the South African Air Force.

Reference table showing RAF serials and the SAAF serials they received. In some occasions, the SAAF registry was already reserved, or maybe even applied prior to shipping.

---

[1]   AS440, AS448/AS451, AS453, AS455, AS459/AS460, AS961/AS962, AS967/AS968, AS971, AS975, AW423 and AW425.

A line-up of Northrop Nomads during a visit to 47 AS Queenstown
(SAAF Museum Collection - Via K. Smy)

An A-17A on a visit to Zwartkop Air Station.
(SAAF Museum Collection - Via K. Smy)

During the first week of January 1942 1267 crashed while on a low flying exercise. The exact date is not clear: this accident occurred on the 6th or 8th. 2/Lt G Niesewand was brought to a hospital but died there on 10 January 1942. The observer, 2/Lt JW Bell, survived the accident.
(SAAF Museum Collection - Via K. Smy)

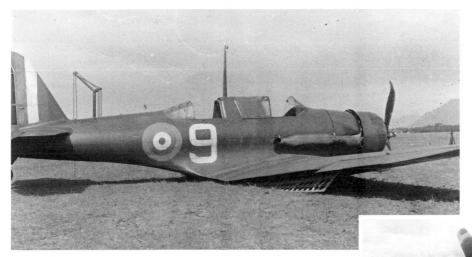

Left and below:
A small number were sent to No. 65 Air School and No. 66 Air School at Youngsfield, Cape-town. 1264, while at No. 64 Air School, was belly landed after the pilot forgot to lower the landing gear. This incident occurred on 27 April 1942. Very little is known about Northrop service at No. 65 Air School.

(SAAF Museum Collection - Via K. Smy)

In July 1942, 35 aircraft remained in service, two more were in storage awaiting transfer to an Air School. Six aircraft had been written off since April 1941. Serviceability of aircraft was influenced by lack of spare parts for the engines. Engine overhaul was carried out by No. 7 Air Depot, but was seriously hindered by lack of spare parts. 14 Engines were stored awaiting overhaul since July. Although the airframes themselves were still in good condition, serviceability figures of the month July were very poor:

No. 41 Air School:	4	23%
No. 42 Air School:	6	83%
No. 43 Air School:	23	38%
No. 66 Air School:	2	unknown

The aircraft were issued to No. 25 Group, consisting of the following Air Schools; The No. 41 Air School at East London received the first samples in March 1941. No. 42 Air School received 7 aircraft on 5 March 1941. *1232* was collected at Cape-town on 10 March. Four machines were transferred to Pretoria and were modified to carry target towing equipment. The first machine was finished on 25 April. Lt. Knowles was sent to Pretoria to test the new equipment. *1246* Crashed near the airfield on 7 june 1941. Crew P/O R.F. Featherstone (RAF) and AC2 N.W. North (RAF) were both killed.

No. 41 Air School got acquainted with the Northrop on 27 March, when a machine visited the School for the first time. Not long after No. 41 Air School received their equipment, a plane was lost. On 6 June 1941 *1244* crashed during a gunnery exercise near East London. Crew R.G. Mur-doch and Sgt. A.J. Barry were both killed. While at No. 41 Air School, Northrop no. *1254* was damaged during landing on 4 July 1941. One of the landing gear legs was not locked properly. It collapsed and the starboard wing was damaged. The crew could escape with cuts and bruises. A similar accident occurred on the 28th of the same month when the same machine was damaged. The pilot had neglected to lower the landing gear. It turned out, that there was no proper Handling Note for the Northrop available. This was quickly prepared and issued to the units. By 31 December, the number of Northrops at 41 A.S. had risen to 14. Four aircraft, however, were transferred to No. 43 Air School during January 1942.

A large number of machines was assigned to No. 43 Air School at Port Alfred, the first arriving there on 15 October 1941. Three more arrived on 3 January, and on the 7th, 8th, 13th, 14th one aircraft each. In the meantime, the Air School was officially activated on the 12th. On 25 April a mobilisation exercise was announced at 12:00 hrs. All eight Northrops were to scramble in squadron formation with Grant's Valley as destination. The flight and dispersal of the aircraft was completed at 13:00 hrs. On the return flight, *1245* of "A" flight suffered engine trouble and made a successful forced landing. It was flown to Port Alfred later that day.

A ninth Northrop was added to the School on 7 May '42, plus two more on the 21st. At this stage, obtaining any materials for the Pratt & Whittney engines was becoming troublesome. Two aircraft were out of commission and stored at Germiston, awaiting new engines.
Northrop *1225* was transferred from East London on 22 June. Two more followed on 24 June. *1233* arrived from Port Elizabeth, and *1251* from East London on the 26th. The spare parts pool was worrisome. Obtaining any materials for the Northrops proved to be practically impossible. Of the 16 aircraft, three were used as parts source

15 April 1942. 1250 hit a tractor whilst taxiing. The crew, Sgt/Pilot RH Durham (RAF), LAC JR Parnham (RAF) escaped without injury. The tractor was operated by 41 Air School. Perhaps the aircraft payed a visit to East London.
(SAAF Museum Collection - Via K. Smy)

to keep the other machines air worthy. Seven aircraft needed an engine refit, while problems with excessive oil consumption was grounding a lot of the remaining aircraft. New piston rings were needed to counter the problems. One engine was completely revised, but after only 5.55 flying hours oil consumption was unacceptably high again.

A single plane was taken on strength on 14 July. *1247* was taken over from No. 41 Air School and arrived on 28 July. Total Northrop strength had now reached 23 machines by the end of July. [1]

1231 arrived from No. 65 Air School at Youngsfield on 13 August. At the end of September 25 Northrops were on strength, the highest number recorded.

---

1   Not all aircraft movements were traceable.

Three aircraft were fitted with drogue towing equipment. Eight aircraft were unserviceable and were stored without engine.

On 30 October 11 aircraft (*1221, 1228, 1231, 1233, 1234, 1235, 1247, 1250, 1251, 1256, 1263*) were airworthy and were transferred to No. 42 Air School (receiving four) and No. 64 (Electrical and Wireless) Air School located at Tempe. 5 More (*1230, 1245, 1257, 1223, 1254*) were transferred to No. 42 Air School on 5 November, plus a single one on the 8th (*1243*) and 20th (*1268*).

At the end of November 1942, seven Northrops remained at No. 43 Air School, although none were airworthy. Two aircraft were reduced to spares and were transferred to Electrical and Wireless school at Bloemfontein. The last engineless machines (*1222, 1225, 1253, 1255* and *1259*) were transported by rail to "various destinations" on 16 December 1942. Thus ended the use of the Northrop at No. 43 Air School. The Northrop had not been involved in a single major accident during its entire period at No. 43 Air School!

At No. 42 Air School the Northrop shared drogue towing duties with Fairey Battles. The only available drogue available, however, gave constant headaches. The drogue was too heavy for the towing equipment of the Northrops, resulting in continuous break downs. A lot of tinkering was nee-

Northrop Nomad 1 "D6" on a visit from one of the 25 group Schools to Zwartkop Air Station.
(Coll. D. Becker)

Northrop Nomad TT 1266 (D17) at Wingfield, near Cape Town, in November 1941. It served as a target tug and was painted in highly visible yellow with black diagonal bands.

(Coll. G. Wood)

ded to achieve a favourable result. The resulting experimental towing equipment was built into Northrop *1240* at Maintenance Command, after which trials were flown. The results were satisfying and all towing equipment would be modified accordingly. The following month, the drogues themselves were modified, after time consuming trial and error.

On 5 January 1943 *1256* of No. 42 Air School stalled and spun into the ground from 1000 ft while drogue towing. Both occupants, lt. E.T. Cowlin and cpl L. Jenkins (RAF), were killed in the incident. During March 11 Northrops were at No. 42 Air School, total flying hours was 131 hours. One machine was out of commission following a gear collapse on 13 March. A serviceability report, dated May 1943, lists only 12 airworthy aircraft remaining.

All remaining aircraft were now concentrated at 42nd Air School. During October the Northrops would take over all drogue towing tasks from the last surviving Battles.

Very little is known about the following period. After November '43 no entries are made in the War Diaries, although it is suggested that they continued to operate up to 1944.

Below:
Northrop Nomads from No. 42 Air School. Leading is 1243 (ex AS958) which was delivered to 42 AS on the SS "Amerika" on 7 Feb 1941. Next is 1226 (ex AW 422) and last is 1265 (ex AS462) coded "15" used first by No. 41 Air School then by No. 21 Air School, and finally put in storage at No. 7 Air Depot on 4 April 1944. (via D. Becker)

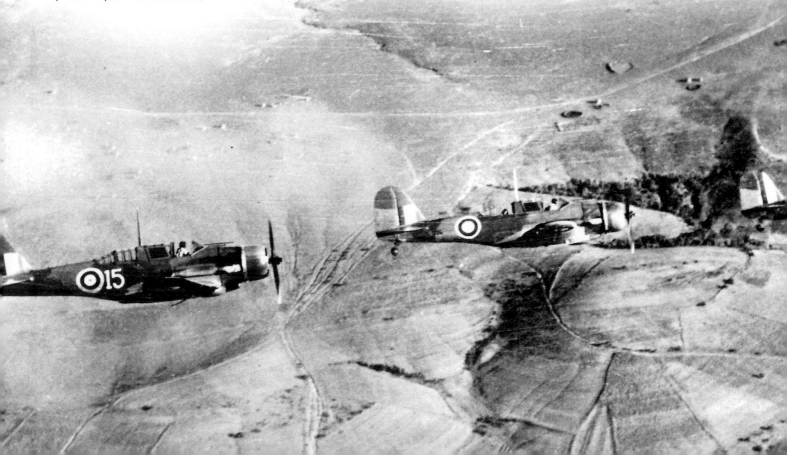

When 'B' Flight of 42 Air School was disbanded in 1944 the unit's Northrop A-17-a Nomads were grounded. It was decided to use one of those for fund-raising purposes, and it was therefore taken to the Donkin Reserve aboard a large transporter and parked next to the lighthouse. The public was then given the opportunity to buy special sixpenny stamps, with the objective of covering the aeroplane with these.
(Painting by R. Belling)

A number of aircraft received a radio call sign on the fuselage. D6, D17 and D18 are known. Here D17 is shown, before it was rebuilt to target tug configuration. (Profile by Alexandre Guedes)

CANADA

The first of 32 A-17A's was accepted at Camp Borden on 8 August 1940. Delivery was completed on the 26th of the same month.They received registries *3490* to *3521*.

The aircraft were used at No.1 Service Flying Training School (SFTS), which was formed at Camp Borden on 1 September 1939. Its initial task was the evaluation and further training of volunteering civil pilots. Despite the relatively inexperienced students, No.1 SFTS had a good safety record while flying the Nomad. Only two major accidents occurred during operations. *3491* crashed and burnt out completely on 19 November 1940. After enquiry, incorrect handling of throttle control was found as cause of the crash.

3505 was taken into service on 13 August 1940. It appears to be cleaned thoroughly – perhaps before receiving a coat of paint?
(Coll. Canadian Armed Forces, Directorate of History and heritage)

3503 did not return from a solo training flight on 12 December 1940.

The following day, Friday the 13th, 50 aircraft were sent out on a search and rescue mission to find the missing aircraft. Two aircraft, *3521* and *3512*, of No.9 Bombing and Gunnery School took off to participate in the search. Disaster struck however, and the two planes collided with each other in a snow storm and crashed in Lake Muskoka. All four crewmembers were killed.

The missing *3503* and its pilot, LAC Clayton Peder Hopton, was found in a swamp the next day, just five miles south east of Camp Borden.

3508 could be repaired after a ground collision.
(Coll. Canadian Armed Forces, Directorate of History and heritage)

Below: *The aircraft received a paint scheme of yellow and black diagonal bands which made them clearly visible when towing target drogues.* *(Coll. Canadian Armed Forces, Directorate of History and heritage)*

3519 as it appeared as target tug, after May 1941. (Profile by Alexandre Guedes)

Nomad *3512* was found in January 1941, and the body of LAC William Gosling could be recovered. The remains of sergeant Lionel Francis, an RAF instructor were found the following June. The other aircraft was not found until July 2010. The remains of LAC Ted Bates and Flight Lieutenant Peter Campbell were recovered in October 2012. The aircraft itself was raised from the lake in October 2014 and is expected to be fully restored.

The availability of spare parts for both aircraft and engines was a major concern. Already during March 1941 this influenced serviceability of the Nomads. It was considered unlikely that spares could be obtained from the United States. It was agreed, that Nomads would be converted to target tug, to supplement the Fairey Battle fleet. Although work on this conversion started slowly, nine aircraft had been modified during May, and another eight would be completed later on.

The Nomads were assigned to Bombing & Gunnery School (B&GS) units and continued to labour on until March 1945.

Of these, No. 4 B&GS operated the Nomads almost without trouble. The only recorded incident is a wheels up landing, made on 20 April 1942, after which the aircraft was repaired. At No. 6 B&GS a number of landing incidents occurred, but all without serious consequence. The Nomad had an eventful time with No. 9 B&GS. Two aircraft were involved in the Friday the 13th incident, which was described earlier. *3506* Was written off on 30 November 1942 when it slammed into fences and a pile of rocks during a forced landing. During 1943 various accidents occurred; On 19 January *3508* was hit by a Fairey Battle, after this plane jumped over the chocks during engine warm up.

On 29 January *3496* got caught in turbulence of a departing aircraft and struck ground with the port wing, which could be repaired. *3504* crashed and was written off after a mid-air collision with a Fairey Battle on 27 March. The planes were participating gunnery practice. *3495* was damaged in the tail after *3502* ran into it after landing on 5 May 1943. *3498* was lost after a mid-air engine fire on 20 July 1943. *3500* Was hit by a landing Fairey Battle on the same day, but could be repaired. A final incident took place on 9 May 1944. *3513* Crashed two miles south-west of Mont-Joli on 9 May 1944 when the engine caught fire in flight while on a target towing exercise.

Other units which have flown with the Nomad were No. 31 SFTS at Kingston, Ontario, and the Air Armament School at RCAF Station Mountain View, Ontario.

A large two digit registry in white replaced the four digit registry on the bottom of the wing. The first three numbers of the old registry is still vaguely visible, 350?.

EXPORT SUCCESS

By Edwin Hoogschagen

Successes of earlier Northrop aircraft attracted foreign interest for the military derivatives of the Gamma family. The first foreign customer was Sweden, which purchased two samples known under the (Douglas) export type designation 8A-1. A further 102 aircraft were produced under license in Sweden. A Northrop Gamma 5B demonstrated in Argentina during June 1936 and this resulted in an order for thirty machines, which were delivered between February and May 1938. Peru was the third customer and purchased, after an earlier purchase attempt was denied by US Congress, ten machines which

were produced as 8A-3P (Peru). The Netherlands signed a contract for eighteen aircraft on 14 March 1939. These were designated 8A-3N (Netherlands). Fifteen aircraft, type 8A-4, were built for the Iraqi air force and were delivered between April and June 1940. The final export customer was Norway, which ordered 36 machines. These were known as 8A-5 but were only ready after Norway had been occupied by Nazi Germany. Instead, the aircraft were diverted to a training camp in Canada known as Little Norway. Thirteen of these were eventually sold to Peru during 1943.

In all, 453 aircraft were built in nearly six years. Three countries used the type in combat and the final aircraft was retired during early 1958, nearly 25 years after the very first YA-13 – the type which ignited development of the A-17 - was built.

Production overview

Variant	c/n	serials	built	Ordered	Delivered	Remarks
A-17	44	35-51	1		27-07-1935	Prototype
A-17	75-183	35-52 – 160	109	01-03-1935	15-01-1936 – 05-01-1937	US air corps
A-17A	189-288	36-162 – 261	100		04-02-1937 – 21-12-1937	US air corps
A-17AS	289, 290	36-349, 36-350	2		17-07-1936, 12-07-1936	US air corps
8A-1	378	751	1	16-04-1937	22-04-1938	Sweden
8A-2	348-377	A-401 – 430	30		22-02-1938 – 17-05-1938	Argentina
A-17A	381-409	38-327 – 355	29	09-11-1937	10-06-1938 – 31-08-1938	US air corps
8A-1	410	7002	1		08-08-1938	Sweden
8A-3P	412-421	XXXI-1 - 10	10		nov 1938 – feb 1939	Peru
8A-3N	531-548	381 – 398	18	14-03-1939	24-08-1939 – 16-11-1939	Netherlands
8A-4	613-627	?	15	?	apr – jun 1940	Iraq
8A-5	715-750	301 – 336	36		nov 1940	Norway
8A-1		7002 – 7041	40	27-09-1938	03-04-1940 – 07-02-1941	Sweden
8A-1		7042 – 7065	24	31-08-1939	31-01-1941 – 20-05-1941	Sweden
8A-1		7066 – 7103	38	11-05-1940	24-05-1941 – 15-11-1941	Sweden

SWEDEN

For a long period Flygvapnet (the Swedish air force) had to make do with outdated material and tactics. Despite ambitious plans and continuous purchases of aircraft, the situation had not improved by the mid-thirties. New plans, dated 1936, were made to modernise and enlarge the air force. This plan was to be completed by the year 1942. Sweden had a modest aircraft industry, and this was unable to turn out a truly modern aircraft. As a result, committees visited various aircraft companies around the globe. A first result was the purchase of a series of German Junkers Ju 86K all metal bombers, followed by the license to produce this type within Sweden's own borders. Additionally, a replacement was needed for the just recently purchased Hawker Hart. A first batch of four aircraft had been ordered at Hawker. Although the Hart had originally been ordered as reconnaissance aircraft, it was tested for possible use as dive bomber shortly after arrival of the first samples. Testing under operatio-

Northrop photos of the first Swedish B 5.
(Coll. G. Balzer)

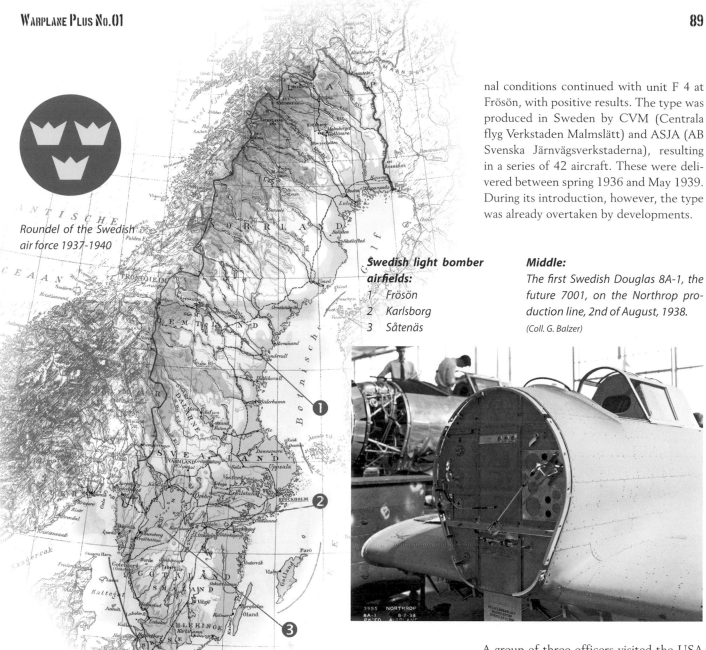

nal conditions continued with unit F 4 at Frösön, with positive results. The type was produced in Sweden by CVM (Centrala flyg Verkstaden Malmslätt) and ASJA (AB Svenska Järnvägsverkstaderna), resulting in a series of 42 aircraft. These were delivered between spring 1936 and May 1939. During its introduction, however, the type was already overtaken by developments.

Roundel of the Swedish air force 1937-1940

Swedish light bomber airfields:
1 *Frösön*
2 *Karlsborg*
3 *Såtenäs*

Middle:
The first Swedish Douglas 8A-1, the future 7001, on the Northrop production line, 2nd of August, 1938.
(Coll. G. Balzer)

A group of three officers visited the USA during March and April 1936. Their objective was to study modern combat aircraft which suited Swedish demands. They visited the 7th Bombardment Group at Hamilton Field and flew the Martin B-10 and Douglas B-18. The North American BT-9 was test flown at Randolph Field. The men finally travelled to Barksdale Field, where they were introduced to the Northrop A-17 which was just entering service. During the same period, a group of engineers from ASJA toured the United States and visited several aircraft companies. This resulted in selecting the Seversky P-35 as new fighter type.

Northrop photos of the first Swedish B 5.
(Coll. G. Balzer)

After returning to Sweden the combined committees advised to buy the rights to build both BT-9 and A-17 types in Sweden. For the A-17, it was recommended to replace the Pratt & Whitney engines with the British Bristol Mercury radial engines. Although its power output was lower, it was feared that the original American engines could fall under an export embargo. The contract was signed on 16 April 1937 and Northrop was given an order to produce two pattern aircraft. The Swedish aircraft were known as Northrop 8A-1. Northrop also offered its services to educate Swedish technicians and to supply technical drawings during the production start-up.

The first Swedish Douglas 8A-1, the future 7001, on the Northrop production line, 2nd of August, 1938. (Coll. G. Balzer)

The first B 5, landing at Malmen, Linköping, in 1938. (Coll. C.G. Ahremark via M. Forslund)

The first B 5 (air force no. 7001) at Malmen, Linköping, in 1938. (Coll. C.G. Ahremark via M. Forslund)

The first machine (c/n 378) was loaded aboard M/S Gripsholm on 22 April 1938 and arrived in Gothenburg in the beginning of May 1938. The aircraft was assembled at CVM, Linköping, and received type designation B 5 and air force no. *751* (later changed to *7001*). It was at first equipped with a Bristol Pegasus XII engine (875 hp). This was later replaced with a Bristol Mercury XXIV engine of 980 hp. This engine type was produced under license in Sweden by NOHAB. This company also made copies of the Hamilton Standard variable pitch propellers.

Before the first aircraft had arrived, ASJA and the technical department of Flygvapnet started negotiations about a first license contract. It was agreed that the aircraft would be slightly modified to Swedish demands; an instrument panel would be added in the rear cockpit and the electrical systems would be altered to Swedish standards. The pilots cockpit featured a heightened canopy and the radio antenna was moved forward to the nose of the aircraft. The first contract for delivery of 40 aircraft, type B 5B, was signed on 27 September 1938. This number included the pattern aircraft *7002*, which would be arriving by ship shortly afterwards. This plane was delivered unassembled and without engine and arrived on 8 August 1938. The parts were used by ASJA for measurements, tooling and as patterns to start up manufacturing. The first aircraft should be delivered on 1 August 1939 and

Skies being fitted on no. 7001 for tests on Lake Roxen. The photo is probably taken in early 1939.
(Coll. C.G. Ahremark via M. Forslund)

one completed aircraft should be delivered every third week. Total cost for the series was 5 131 155 SEK (Swedish Crowns). Six of the aircraft were to be equipped with dual controls. ASJA had received another substantial order slightly earlier, to produce 35 North American BT-9s. The first of these would enter Flygvapnet service during May 1939 as Sk 14.

Delivery of the B 5Bs suffered delays. Acquiring the needed raw materials proved to be a difficult undertaking. A half year after the contract had been signed, it was estimated that the 40 aircraft could be delivered between 1 January 1940 and 15 January 1941, slightly behind original planning. With war threatening Europe, it was decided to buy 50 complete undercarriage sets directly from the USA. To accelerate the preparations, a Swedish company was hired to produce the necessary jigs and molds.

Despite start-up problems, 24 extra machines were ordered from SAAB on 31 August 1939. An order for 25 machines to be built by Northrop was contemplated, but this option was not pursued. After war had erupted on 1 September, delivery of the landing gears from the U.S. was no longer possible, and SAAB copied and produced these themselves. The radio direction-finding station was later copied by Aga-Baltic.

The first B 5B, no. 7002, which was the pattern aircraft for production, was delivered to Flygvapnet on 3 April 1940. Aircraft nos. 7003 and 7004 were also delivered during April. By the end of June, eight aircraft had been delivered. Unfortunately, no. 7011 was destroyed in a fatal crash on 21 September, just a few weeks after delivery. When recovering from a aerobatic manoeuvre, the outer wing panel broke

off, taking with it part of the tail plane. The crash was investigated thoroughly and as a result, the wings of all aircraft needed to be strengthened. Construction of a large number was already well under way and these had the be modified on the production line. The last machine was finally delivered on 8 February 1941. The first aircraft of the second series had been delivered during the first months of 1941 as well, and the final machine was handed over on 20 May of that year.

On 11 May 1940, a final series of 38 aircraft (B 5C) was ordered from SAAB (the order had been preceded by a discussion to order 160 aircraft). The situation in the Scandinavian sub-continent had significantly changed in the previous months. Finland had been involved in a bloody conflict with the Soviet Union and on 9

April Germany had assaulted neighbouring countries Denmark and Norway. And on 10 May, Germany launched its offensive against Holland, Belgium and France.

The manufacturing of B 5s at Saab on 23 April 1941.

(Karl Sandels Svenskt Fotoreportage, from P. Haventon's archive, via M. Forslund)

Technical data	
Engine	Bristol Mercury XXIV
Power	980 hp
Wingspan	47 ft 9 in (14,55 m)
Length	31 ft 8 in (9,7 m)
Height	12,3 ft (3,76 m)
Empty weight	5,368 lb (2.435 kg)
All-up weight	9,369 lb (4.250 kg)
Max. speed	208 m/ph (335 km/h) at 2.000 m
Range	497 m (800 km)
Service ceiling	22,635 ft (6.900 m)
Climb rate	-
Armament	4 fixed 8 mm, one flexible 8 mm mg
Bombload	maximum 800 kg

The B 5B was able to carry a number of bomb load combinations, with a maximum of 700 kg. A 500 kg bomb could be carried under a swing-down bomb rack, quite similar as the central bomb rack fitted on Junkers Ju 87 Stuka and Douglas SBD Dauntless. Alternatively, a 250 kg bomb could be carried under the fuselage. Four 50 kg bombs could be carried under the outer wing panels.

Assembly of engines for the B 5 at SAAB, Linköping

(Karl Sandels Svenskt Fotoreportage, from P. Haventon's archive, via M. Forslund)

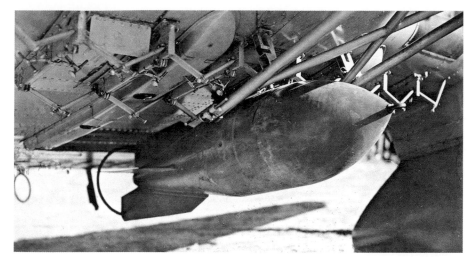

controls. Construction started during May 1940. The first machine, *7066*, was delivered on 24 May 1941 and the final one of the series on 15 November of the same year. In all, 103 aircraft had been delivered. 19 of them were fitted with dual controls. After delivery, aircraft received a lengthened exhaust.

A 250 kg bomb attached to the swing down bomb rack. (Armé-, Marin- och Flygfilm, via M. Forslund)

The bombload increased to 838 kg on the B 5C, which was able to carry 24 12 kg bombs internally and 11 50 kg bombs on external racks. With this bomb load combat range of the aircraft dropped to just 113 km. Another option was a bombload containing six 50 kg bombs attached vertically in the internal bomb bay, leaving external racks available for other ordnance.

Furthermore, the B 5C featured an enlarged fuel capacity and during production, improved engine cowlings were fitted which featured adjustable cowl flaps. Six of the machines were fitted with dual

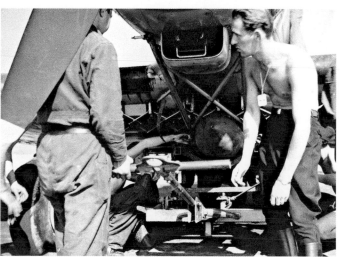

A 250 kg bomb being loaded on a B 5 from F 4. (Armé-, Marin- och Flygfilm, via M. Forslund)

(Svensk Flyghistorisk Förening via M. Forslund)

INTO SERVICE

The B 5 entered service with F 4, based at Frösön. During summer and fall of 1940 crews followed an intensive training scheme to familiarise them with their new planes. Meanwhile, F 4 continued to fly a large number of B 4s. Delivery to the second light bombardment unit, F 6 at Karlsborg, was delayed, therefore training started with three Sk 14s and six B 4s. A combined exercise took place between 28 August and 16 September. A cross country flight, bombing sorties and fighter interceptions were flown. First experience with bombing were exercises with 8 kg concrete bombs and light 2 kg bombs. During the exercise valuable experiences were gained with large formations. Transition from the B 4 to B 5 proved to be big. The B 5 was a truly modern aircraft, featuring flaps, constant speed propeller, radio direction finding and a large amount of instrument equipment.

Getting familiar with the B 5 had its ups and downs. Here, B 5 no. 7003 after an accident during landing, 16 May 1940. (via Krigsarkivet)

Some combat ready B 5s in the early 1940s. The nearest machine carries a 250 kg bomb on the centre bomb rack and two 50 kg bombs under each wing. (Armé-, Marin- och Flygfilm, via M. Forslund)

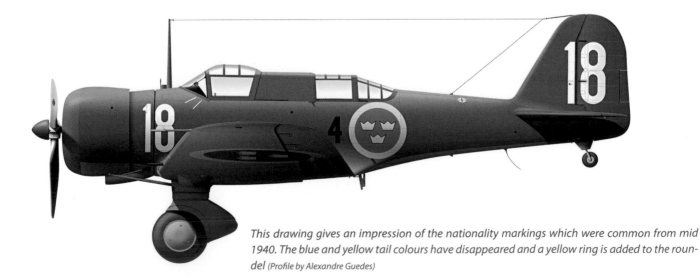

This drawing gives an impression of the nationality markings which were common from mid 1940. The blue and yellow tail colours have disappeared and a yellow ring is added to the roundel (Profile by Alexandre Guedes)

units and army units. During these exercises, some incidents occurred. One aircraft had to be force landed due to ice building up on wings and tail planes. Another crew got lost and made a safe landing some 15 kms removed from Luleå.

The M-41 sight.

(via Krigsarkivet)

In the end of 1940, aircraft no. *7030* (SAAB c/n 164) was equipped with Edo 7850 floats at Torslanda, Gothenburg.

F 6 received its first B 5s during December 1940. These were actually transferred from F 4. Each unit was planned to operate 36 first line aircraft, divided over three units (division) of twelve aircraft each. A small number of machines were attached to staff units. During delivery, F 4 continued to fly the B 4 up to the final months of 1941.

During the 1941 winter exercises F 4 was able to send a division up north, where several units had gathered on the frozen waters surrounding Luleå. From here, patrol flights along the border with Finland were undertaken. The B 5s were available for dive bombing duties and operated with ski undercarriages. Flying qualities were hardly effected when the machines were fitted with skis.
Starting February 1941, bombing techniques were studied. Various bomb loads were tested, at dive angles from 45 to 55 degrees. Dive brakes were used to keep dive speed within the critical speed of 440 km/hr and the limit of 4,5 G. At first, a simple ring and bead sight was used for bomb aiming and firing the fixed machine-

guns. This was fitted in front of the wind screen.

F 6 held large scale exercises during the whole of February 1941. The unit used a frozen lake near Lulea as landing strip. Skis were fitted to operate in snowy conditions. Flights were made with other Flygvapnet

These resulted in a weight increase of 213 kg. During the tests the aircraft was designated B 5S.

Rudders were added to the floats, to improve manoeuvrability on water. The aircraft suffered severely during the tests; salt water took a heavy toll on the aircraft.

A movable 8 mm machine gun was placed in the rear cockpit, seen here on a B 5 from F 4 on 5 september 1941. (Svensk Flyghistorisk Förening, via M. Forslund)

Corrosion was found between engine and firewall and in the rear part of the wings. Cracks in the propeller blades were found on 23 September, and after a test flight in November 1940, it was found that the aircraft had deteriorated too much to safely continue the tests. Among other things, cracks in the floats had been found, which were caused by many hard landings.

During spring of 1941 the number of delivered machines had increased steadily and now, subscripted pilots, wireless operators and gunners were undergoing training on the B 5, and during July, all six divisions of F 4 and F 6 had received their full inventory of aircraft. The B 4s were now transferred to the first division of newly formed F 7 stationed at Såtenäs. It were tense times. Germany had launched its attack against the Soviet Union. F 4 and F 6 were drawn closer to the border with Finland and frequently operated from forward airfields. Exercises were a continuous part of the daily routine. F 4 loaned four machines to F 7, where the student pilots were preparing for flying with SAAB B 17 aircraft. More large scale exercises were held during summer of 1941.

During the summer of 1941 the movie "Första Divisionen" was filmed. F 6, at Karlsborg, provided the scenery and many flying hours.

(Photo from F 6's archive, via M. Forslund)

F 6 operated from Getterön airfield near Varberg and dispersal field 10 near Enstaberg. Dispersal fields were often simple grass strips, with wood barracks and a few simple hangars. Most of the maintenance had to be improvised. Filling up the fuel tanks with the hand pump, 1100 liter capacity in total, could be a long and tiresome job.

F 6 lost two B 5s in a mid-air collision on 12 August 1941. Aircraft *7068* and *7069* collided during a formation flight and both crashed directly afterwards, one going down in a lake and the second crashed on the beach.

Despite international tensions, F 6 provided the backdrop and many flying hours during the filming of the Swedish drama movie "Första Divisionen", which premiered during September 1941.

B 5s from F 4 on skies. The winter of 1941-42 was particularly fierce but operational flying continued nonetheless.

(Armé-, Marin- och Flygfilm, via M. Forslund)

Disregarding the ruthless winter weather, flying continued during January and February 1942. Flygvapnet was put on high alert when embargoed Norwegian shipping was subject of an diplomatic dispute with Great Britain and again when tensions with Germany grew during February. F 6 was ordered to move to Hammerdal. Flygvapnet also served local residents in the north. During the icy conditions B 5s from F 6 dropped canisters containing food and supplies to isolated villages. When thaw set in, Flygvapnet finally rotated its units and the B 5 equipped units were releaved.

2. Flygeskadern comprised F 7 which operated the SAAB B 17, together with fighter units F 9 and F 12. F 3 and F 11 also provided one division each for reconnaissance tasks. Other Flygvapnet units were attached to Army or Navy or stood under direct command of the airforce commander. More importantly, new tactics had been developed. Instead of the near vertical dive, a more shallow dive attack proved to be less strainfull for crew and machines.

Enrollment of this new doctrine was the first indication that the role of the B 5 as first line aircraft was to change dramatically.

F 4 and F 6 both took part during the large September exercises and operated from forward airstrips in Västergötland, in the south of Sweden. Fighter aircraft from F 8, F 9 and F 10 acted as enemy forces. Unfortunately, one of the B 5s from F 4 was lost during the exercise, when *7054* was lost on 7 September during a tragic accident. The machine lost a live bomb during take-off, which instantly exploded. Other bombs detonated in the chain reaction. *7054* was consumed by fire and the crew was most likely instantly killed. As a result, B 5s were not allowed to carry live bomb loads during the period of investigations.

Flygvapnet underwent many developments during 1943. New types entered service, and the structure was reorganised. F 4 and F 6 were attached to newly formed 1. Flygeskadern. The first divisions of F 3 and F 11 served in the reconnaissance role.

A ski equipped B 5 is being assisted during taxiing. In the air is a J 11 (Fiat Cr.42) from F 9 with winter camouflage. (Armé-, Marin- och Flygfilm, via M. Forslund)

B 5s from F 4 on skies. The winter of 1941-42 was particularly fierce but operational flying continued nonetheless. (Armé-, Marin- och Flygfilm, via M. Forslund)

A B 5 in front of its successor Saab B 17 and two B 3s (Junkers Ju 86).
(C. Palmblad, via M. Forslund)

The SAAB B 17 was set to replace the B 5 at F 6, starting fall 1943. The process was planned to be complete by the end of 1944. The B 17 was a more modern aircraft, with better handling qualities, flight qualities and more modern equipment. The new m/42 bomb sight was introduced on the B 17, but before it entered service, it was tested on B 5 no. *7025*. Results at dive angles between 30 and 50 degrees were very promising. Although F 4 and F 6 were still front line units, replacement of aircraft was prepared. During December F 6 transferred twelve aircraft to F 4, which was putting airframes in storage. 50 Machines would be modified to serve as target tug and as glider tug. Before this modification started, F 4 continued to fly its neutrality patrols in the north. The unit transferred to Abisko at lake Torneträsk near the Norwegian border and settled down at a forward airstrip near Lulea. This cross country flight was part of training. During this particular flight one of the machines, no. *7004* was reported as missing. The wreckage, and the remains of its crew, was finally found in December 1948.

A B 5 from F 6 with winter camouflage. A badge, in the form of a bull, has been applied. (C. Palmblad, via M. Forslund)

B 5s with winter camouflage. The planes carried various, individual, schemes with many differences.
(Svensk Flyghistorisk Förening, via M. Forslund)

Practice bombs are sleighed to a waiting B 5. This machine has received a winter camouflage with a splinter camouflage variant.
(Coll. E. Hoogschagen)

During the period September 1943 to January 1944 a primitive G suit was tested during dives with a B 5. A small compressor was built into the rear cockpit, which inflated a rubber, rather cumbersome, suit during pulling out of a dive. This concept was later studied after introduction of the first Flygvapnet jets.

After the summer of 1944 the B 5 had completely disappeared from F 4 and F 6 inventory. Both were now operating B 17, as was F 7. During the operational period, spanning November 1940 to July 1944, 23 aircraft were lost in accidents. 31 crewmen were killed.

Mishaps		
21 September 1940	7011	Crashed near Linkoping. Crew Sigvard Lundgren, Per E. Starck killed.
18 February 1941	7047	F 4. Crashed during take-off for night flying. Crew escaped unhurt.
12 August 1941	7068	F 6. Mid-air collision with 7069 near Ljugarn. Crew Lars E. Kock, S. Gustav Hylander killed
12 August 1941	7069	F 6. Mid-air collision with 7068 near Ljugarn. Crew Sigurd Berglund killed, observer Larsson escaped.
19 August 1941	7059	F 6. Crashed after stall at low altitude. Crew escaped unhurt.
25 November 1941	7100	F 4. Hit the ground during low level flight. Crew Gösta L. Nydahl, Ernst A. Dahlin killed.
3 December 1941	7084	F 6. Crashed after a/c failed to recover from dive. Crew F. G. Holmgren, B. Å. Holmdahl wounded.
24 March 1942	7077	F 6. Crashed after engine fire. Crew Åke F. V. Borg, Nils F. Eriksson killed.
19 May 1942	7089	F 6. Crashed near Gårdsjö. Pilot Folke Blomdahl killed.
12 June 1942	7082	F 6. Crashed near Mölltorp after mechanical failure, pilot Björn Lindskog escaped.
16 June 1942	7027	F 4. Crashed during dive bombing exercise. Pilot Karl Rune Sporre killed.
7 July 1942	7034	F 6. Crashed into water during gunnery exercise. Pilot Axel Axelson killed.
28 July 1942	7101	F 6. Crashed after a/c failed to recover from dive. Crew Per O. F. Strandberg, Erik O. L. Berglund killed.

At the end of its career some B 5s were modified for target towing. The equipment is visible bellow the tail. This photo is from Kalix 1946. (H. Ängeskog, via M. Forslund)

After the B 5 was relieved from front line duty the remaining machines were sent to SAAB for modifications. Twenty nine aircraft received blind flying equipment, while 50 machines were re-equipped as target and glider tugs. In this role, the B 5 served with all of Flygvapnets units. On 28 April 1944 a B 5 was towing a transport glider, type Lg 105, which was under development. At a height of 1500 meters heavy vibrations tore off pieces of the gliders' wing, resulting in a fatal crash. The B 5 was flown to safety unscaved. The B 5 continued to serve as target and glider tug up to 1948, when it was again succeeded by the B 17. In the final three years of flying no less then fifteen aircraft were lost. The final two machines were written off on 1 July 1950.

More mishaps		
7 September 1942	7054	F 4. Destroyed when bomb prematurely exploded. Crew Carl Stellan Sigfrid Jufors and Erik Valdemar Lundström killed.
10 February 1943	7073	Entered spin during mock dogfight, crashed near Hisingen, crew Torsten Edvard Grahn, Rune Vilhelm Andreasson killed.
26 March 1943	7038	F 4. Crashed near Sunne. Crew Karl Einar Mellström, Stig Harald Siggstedt killed.
7 May 1943	7083	F 6. Destroyed during crash landing near Askersund. Crew Karl Göte Jansson, Karl Åke Lennart Källman killed.
19 May 1943	7070	F 6. Crashed near Perstorp during bombing exercise. Pilot Åke O. M. Olsson killed.
14 July 1943	7037	F 6. Crashed after midair collision with J 11 fighter plane. Crew Bror Valter Viking Widung and Göte Ragnar Törnfeldt killed.
8 July 1943	7049	F 4. Crashed following engine failure directly after take-off. Crew Erik Wittberg, P. Yngve Östlund killed.
14 February 1944	7004	F 4. Crashed near Överkalix. Crew Erik Agerberg and Bengt Erik Aspgren killed.
13 June 1944	7010	Heavily damaged in forced landing near southern tip of Öland, subsequently written off. Pilot Lundin escaped unhurt.
20 June 1944	7087	F.6. Slammed into parked aircraft after landing at Sjögestad airfield. Crew Lars-Åke A. Smedinger, C. Åke E. Hain killed.

Another view of a target towing B 5. Armament has been removed, while a frame has been placed around the tail wheel to prevent the target drogue tangling up in the tailwheel. (S. Gripenlöf, via M. Forslund).

In the end of the 1940s some B 5s were sent to various flottiljer for training and various tasks. Here is a B 5 from F 13.
(H. Ängeskog, via M. Forslund)

And a B 5 with markings from F 8, Norrköping.
(W. Jansson, via M. Forslund)

F 4 also used letter codes on their B 5s. These replaced the tailnumber and appeared during September 1943.
(Å. Engström, via M. Forslund)

Northrop 8A-2 in Argentina

By Santiago Rivas

By the mid-thirties, the Argentine Military Aviation, dependent from the Army, needed to be re-equipped accordingly, to keep the regional balance, which was lost because of the obsolete equipment they had. Because of this, in 1935 and during the government of president Agustín P. Justo, the need to buy new airplanes for the Army and the Naval Aviation was discussed. By then, the fleet comprised on the FMA Ae.M.Oe.1 for observation and training, FMA Ae.M.E.1 trainer, Dewoitine D-21 and D-25 fighters while the FMA Ae.M.B.2 Bombi and Junkers K43 bombers and transports replaced the old Breguet XIX. The transport fleet also comprised Junkers F13, FMA Ae.T.1 and Ae.M.S.1 and the trainers also included Avro Gosport built under license in Argentina and Avro Trainers.

The lack of power and age of most of them, added to some accidents, determined that, after an intense debate in congress, on August 1935 the law 12,254 was sanctioned, indicating the need to purchase new airplanes and weapons for the Comando de las Fuerzas Aéreas del Ejército (Army Air Forces Command) and law 12,255 did the same for the Servicio de Aviación Naval (Naval Aviation Service) on what must

Roundel of the Fuerza Aérea Argentina, or FAA

Main Argentine airfields

1   El Palomar
2   El Plumerillo
3   Parana
4   Coronel Pringles

429, as it appeared up to August 1944 (Profile by Alexandre Guedes)

be a joint purchase using the funds provided by law 11,266. As an intermediate measure, as there was a need for a modern fighter, ten Curtiss Modelo 68A Hawk III were purchased, which won the competition against the Fiat Cr.30 and the Dewoitine D-371 and D-500.

On 7 April 1936 an international bid was opened for the provision of a determined quantity of planes, which, after several changes finally counted 22 bombers, 30 fighters, 30 attack planes, 3 navigation and bombing trainers, 20 basic trainers, 30 advanced trainers, 2 autogiros, five transports and three for transport and liaison. Meanwhile, on 17 November, the Dirección de Material Aeronáutico ordered Madsen machine guns of 11.35 and 7.65mm to be use on the Ae.M.B.2 and Ae.M.Oe.1.

Offers arrived from a wide array of factories and this included, for the single engine bomber program, the Northrop Gamma, Vultee YA-19, the Breda Ba.65, Bellanca 28-70, Curtiss Model 60, and unknown models from Breguet, Koolhoven (most probably the F.K. 52) and an offer from a company described as Tri American Aviation Inc.

A commission was organized under the command of Colonel Víctor Majó, to check the proposals and then test the planes taken to the country. They later included the purchase of basic and advanced trainers,

A large number of the competing aircraft have been assembled at El Palomar. (Coll. S. Rivas)

transport planes and bombing trainers.

The contenders showed their models at BAM (Base Aérea Militar) El Palomar, near Buenos Aires, which included the IMAM Ro.37 (serial *I-APLA* with a radial engine Piaggio P.XR of 700 HP and nine cylinders), a Northrop Gamma 5B (serial *NR14998*) and the Vultee YA-19 (serial *NR14980*). Tests began on Monday 7 September 1936 at El Palomar, starting on 14 September with the Northrop Gamma and the Vultee YA-19. Early in October, the steamer *Indiam* arrived from Italy with the Ro.37.

All those planes were tested during September 1936 until the end of the year, the Northrop Gamma and the Vultee YA-19 were the ones that generated more interest and in the end, the Gamma was selected, but on the Northrop 8A-2 version, while the YA-19 was later selected by Brazil on its V-11 version.

With a smart criteria, the planes were selected trying to standardize engines and accessories, so the final selection included the Curtiss Hawk 75-O fighter, Glenn Martin 139 WAA as bomber and as attack and light bomber the Northrop 8A-2.

The first plane before its delivery on the United States. (Coll. S. Rivas)

View of the 409 from below, where the dive brakes and bomb racks can be seen. The registries were carried on both wings in large black numbers. (Coll. S. Rivas)

All had almost the same Wright Cyclone engines, which were also installed on the five Junkers Ju 52/3m.

At last, 11.35mm Madsen machine guns were ordered until reaching 104, while 157 Madsen of 7.65 mm were ordered, as all planes were ordered without weapons. Fifteen Fairchild Type K3B Photo cameras were ordered and 75 Telefunken Stat 274AF radios for the planes, among other equipment.

CHARACTERISTICS OF THE ARGENTINE NORTHROP 8AS

The Northrop 8A was a very robust plane, which survived many accidents, with some planes having up to four during their career, being repaired or rebuilt. Some, which had turned over, received a different windscreen and others lost the capacity to use the retractable photo camera.

Like all Argentine combat planes, the aircraft were flown with polished metal surfaces, with black anti-glare panel ahead of the cockpit, big roundels on the wings and smaller ones on the sides of the fuselage and the rudder with the Argentine flag.

Northrop 8A before delivery at Douglas Aircraft on 28-1-38, during tests with the full bomb load of six 55 kilos bombs. (Coll. S. Rivas)

It was the first all metal attack plane of the Argentine Army Aviation but they arrived close to the Second World War and soon they became obsolete. Because of the space in their wings, they were equipped with two 7.65 mm and two 11.35 mm Madsen machine guns, but usually they carried only two of one of those calibers. The guns were loaded with 925 and 710 shots respectively. Another machine gun of 7.65 mm could be placed in the observer position, turning the seat of the observer to the back. This gun could be used to defend the plane and also, firing to the side and below, for strafing ground units. The observer machine gun carried 1,000 shots.

The aircraft were also equipped with two illumination flares, that could be fired at less than 240km/h and from a minimum height of 610 meters, to eliminate the risk of causing a fire with the flares touching the ground still burning.

The aircraft could carry various bomb loads; 20 fragmentation bombs of 13,6 kg (on internal cells in the lower part of the fuselage) plus six general purpose bombs of 55 kg, four of 130 kg, two of 270 kg or one of 550 kg under the central part of the wing. The maximum bomb load was 615 kg.

View from behind of the 401 before its delivery, on 31 January 1938.

(Coll. S. Rivas)

Left:
Ground crew installing 55 kilos bombs.

(Coll. S. Rivas)

Mechanics preparing a Northrop 8A for a gunnery training. The plane is equipped with 7.65mm machine guns. (Photo Time Life).

Left: *A Northrop being refueled at El Plumerillo Air Base. (Photo Time Life)*

A Fairchild K3B photo camera could be fitted in the retractable position on the belly, operated by the observer on a very uncomfortable position inside the fuselage, having four small windows on the sides of the fuselage to have some light and a reference of the terrain. This place was sometimes used to carry a third occupant or to carry some cargo, like spares to a campaign aerodrome.

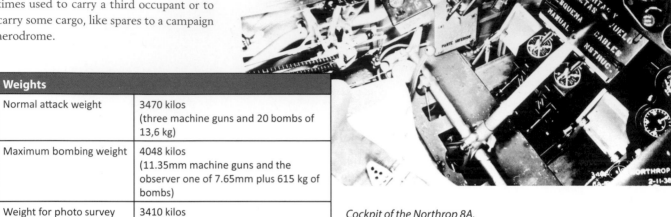

Cockpit of the Northrop 8A.
(Coll. S. Rivas)

Weights	
Normal attack weight	3470 kilos (three machine guns and 20 bombs of 13,6 kg)
Maximum bombing weight	4048 kilos (11.35mm machine guns and the observer one of 7.65mm plus 615 kg of bombs)
Weight for photo survey	3410 kilos (three machine guns and photo camera)

Northrop 8, N.A.16 and Martin 139 WAA at Base Aeronaval Punta Indio in 1938. (Coll. S. Rivas)

Northrop 8 armed with 11.35mm machine guns. (Coll. S. Rivas)

During its career it was also used to deploy smoke screens, receiving special equipment on the ventral bomb rack. Two units were modified to tow aerial targets for air-to-air combat training for the Curtiss Hawk 75-O or the Curtiss Hawk III of Regimiento Aéreo N°2.

They were extensively used for photo survey over the Argentine territory, for the creation of maps by the Military Geograp-hic Institute, other state organizations and for the organization of air routes. When the plane flew without the observer it was necessary to put a weight of 90 kilos on his position, to keep the centre of gravity, as the plane had a tendency to put down its nose while flying. They were equipped with a Fairchild Kruessi RC 3 radio compass which worked very poorly.

The Wright Cyclone R-1820G-3 engine used on the plane had 840 HP at a height of 14,000 feet, working very well at higher altitude, something tested extensively over the Andes mountains of Mendoza Province. For photo survey the best height was 9,000 feet but for reconnaissance they operated at a lower altitude.

Technical data Northtrop 8A - 2	
Engine	Wright Cyclone R-1820 G3
Power	840 HP at 2100 rpm and 14,000 feet
Wingspan	47 ft 9 in (14,55 m)
Length	31 ft 8 in (9,65 m)
Height	12,3 ft (3,76 m)
Empty weight	4,729 lb (2.145 kg)
All-up weight	8,924 lb (4.048 kg)
Max. speed	220 m/ph (355 km/h) at 2.651 m
Range	1180 m (1900 km)
Service ceiling	25,393 ft (7.740 m)
Climb rate	-
Armament	Two fixed Madsen of 7.65mm, and/or two fixed Madsen of 11.35 mm. One Madsen 7.65 mm observer machine gun.
Bombload	615 Kg (maximum).

The 430 on its early days. (Coll. S. Rivas)

Planes at El Palomar. (Coll. S. Rivas)

THE NORTHROPS ARRIVE

On 3 June 1938 the 30 Northrop 8-A were officially enlisted on the Dirección General de Material Aeronáutico del Ejercito, receiving serials *401* to *430*. The first examples had arrived on February of that same year. It's first destination was the Agrupación Instrucción (Instruction Group) Base Aérea Militar (BAM) El Palomar, on the outskirts of Buenos Aires, where the planes serialled *426* to *430* served, together with five Curtiss Hawk 75, six Martin 139 and 15 NA-16 to train the first pilots that were assigned to fly the planes, performing an intense training on the plane which included combat, dictated by the US Military Mission in Argentina, headed by Colonel John Cannon of the USAAC. After the training, 25 Northrop 8-A were sent by the end of 1938 to the Grupo de Bombardeo Liviano of the Regimiento Aéreo N°3 at BAM El Plumerillo, in Mendoza, where they served together with four Ae.M.B.2 Bombi and five Ae.M.Oe.2

When on 24 January 1939 an earthquake devastated the area of Chillán in Chile, three Northrop 8A were sent from Mendoza to Santiago carrying one ton of medicines needed urgently, making the crossing of the Andes cordillera without incidents.

Formation at the Escuela de Aviación Militar in Córdoba, with Northrop 8s together with Curtiss Hawk 75-O, Hawk III, Dewoitine D.21, North American N.A.16 and Focke Wulf Fw 44J. (Photo Time Life.)

View of Pavillion 36 of the FMA (Repairs) where three Northrop 8 could be seen beside a single Martin 139WAA, two Ae.M.B.2 Bombi, one Fw-58 Weihe and three Ae.M.O. Note that the plane closer to the camera has the retractile system for the camera down. 426 was lost on 7 September 1940. (Coll. S. Rivas)

During a training flight, on 25 August, the first accident on the model occurred, when *402* hit a hill on the area of Punta de Vacas, Mendoza, causing the loss of the pilot. Despite this loss, the adaptation to the new material was fast and by the end of 1939 the situation of BAM El Plumerillo was with planes *401, 405, 409, 410, 413, 418, 420, 422* and *424* in service. The *402* and *412* accidented, the *403, 404, 408, 415* and *417* on repairs, the *419* and *425* on commission with other units, the *404* and *406* on commission at BAM El Palomar, the *407* and *421* on commission at the Escuela de Aplicación at the Fábrica

Left:
The 421 at El Palomar.
(Coll. S. Rivas)

Below: *The plane 423 at El Palomar with a special color stripe applied for an exercise in 1941. (Coll. S. Rivas)*

Militar de Aviones (FMA, Military Aircraft Factory), the *414* was on repairs at the FMA and the planes *407*, *411*, *416* and *423* were without engine.

In March 1941, seven planes, with serials *401*, *403*, *410*, *417*, *420*, *422* and *425* were destined to the Escuela de Aviación Militar (Military Aviation School) at Cordoba, which was dependant to the Centro de Instrucción de Aviación (created by the end of 1939), where they served together with three Curtiss Hawk 75-O, one Focke Wulf FW 58 Weihe, 26 North American NA16-1P and other trainers. Also, fifteen Northrop 8A were sent, together with fifteen Curtiss Hawk 75-O to the Regimiento Aéreo Escuela, at BAM El Palomar, also subordinated to the Centro de Instrucción de Aviación, leaving the other seven planes at Mendoza.

Shortly after, in April, the plane serialled *414* was sent to the Grupo de Observación N°1, at BAM General Urquiza, Paraná, Entre Ríos province, leaving on the Regimiento Aéreo N°3 at BAM El Plumerillo a force of eleven Ae.M.B. 2 Bombi, seven Ae.M.Oe.2 and the Northrop serials *405*, *411*, *412*, *413*, *415* and *418*. Finally, the *422* never went to the Military Aviation School, as it was destroyed on an accident at El Plumerillo on 2 April.

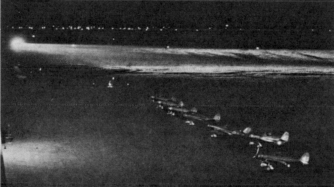

Northrop with some N.A.16 at BAM Villa Mercedes.
(Coll. S. Rivas)

Middle: *Northrop 8A and Curtiss Hawk 75-O during night operations training at El Palomar.*
(Coll. S. Rivas)

The plane serialled 423 at El Palomar. Note the color stripe over the serial on the side of the fuselage. (Coll. S. Rivas)

The plane 429 at El Palomar in 1942. Note the stripe on the fuselage and the wings used for exercises. (Coll. S. Rivas)

During the year and by Orden Reservada N°16 (Reserved Order) of 7 August 1941 the planes serialled *407*, *408* and *409* of the Regimiento Aéreo Escuela were transferred to the Grupo de Observación N° 1.

By those times, the new airplanes led to many changes on the organization of the force and less than one year later it was changed again, when on 16 January 1942, by Orden Reservada N°27, the Commander of Army Aviation ordered that "the planes Glenn Martin, Northrop and Curtiss 75-O which are at the Escuela Militar de Aviación will be sent immediately to their unit of origin, on the same way they were delivered, including the mechanics and equipment. The flying units will be organized on the following way: Regimiento Aéreo N°1 with nineteen Glenn Martin; Regimiento Aéreo N°2 with 42 Curtiss Hawk 75-O and three Curtiss Hawk III; Regimiento Aéreo N°3 (Mendoza) with 24 Northrop 8A-2, including the *414* which belonged to the Grupo de Observación N°1; Agrupación Entrenamiento Paraná, with the three Northrop serialled *407*, *408* and *409*, among other planes".

As an experiment to test new camouflage paint and for a short period, the plane *406* received a new scheme during 1942, with the upper parts in two tones of green and the lower one in light blue. This paint replaced the polished aluminum used until then, with the Argentine flag on the rudder and no serials. The plane suffered a minor accident at La Paz, Mendoza, on 23 November of that year and was repaired later.

Later, on 27 January 1943 the plane serialled *405* was destroyed at El Algarrobal, Mendoza, on what was its third accident, but with no victims. Shortly after, on 4 February, another accident took place, this time at Real de la Cruz, also at Mendoza, when the plane *423* was lost and its two crew members were killed.

New paint scheme

After the tests were completed, a new Reserved Order dated 7 October 1943 stated that the paint scheme in use was to be changed and "the Dirección General de Material Aeronáutico will proceed, urgently, to paint the planes with a scheme like the Beechcraft C-45A used by the US Aeronautical Attaché in Buenos Aires".

From then on, all combat planes received dark green on the upper part and light blue on the lower surfaces, keeping the roundels on the same positions and the Argentine flag on the tail in smaller size. The first plane to receive the new paint scheme was the Curtiss Hawk 75-O serial *630*. The new paint wasn't applied with a primer so after some time most planes started to lose it, looking as if they were in very bad conditions.

Most of the accidents on those times occurred in Mendoza, like the one happened to the *424* near BAM El Plumerillo, on 1 August 1944, when the plane capsized while taking off and burned, injuring the pilot. By the end of the year the *420* also suffered an accident, but with no victims.

The A-406 wearing an experimental paint scheme on three tones of green. (Coll. S. Rivas)

The A-403 with the paint in bad conditions clearly seen. The Vehicle in front was equipped to start the engines of the planes. *(Coll. S. Rivas)*

A strengthened wind screen was mounted on aircraft A-412.

SERIAL CHANGE

On 26 August 1944, the Orden Reservada N°15 was issued with the aim of avoiding the problems caused by the existence of different plane models with the same serials, deciding to add a word which indicated the mission of the planes and states that "the Aeronautic Commander in Chief ordered that the planes in service at the Comando de las Fuerzas Aéreas Militares (Military Air Forces Command) and the Dirección de Institutos Aeronáuticos (Aeronautical Institutes Direction) will receive one or more words on their serial to indicate its mission" and included A-Ataque (Attack) for the Northrop 8A-2, B- Bombardero (Bomber) for the Glenn Martin 139WAA and C- Caza (Fighter)

The A-414 *at Mendoza. (Coll. S. Rivas)*

A plane at BAM Los Tamarindos on their last times at Mendoza. (Coll. S. Rivas)

for the Curtiss Hawk 75-O. Since then, the Northrops become serialled *A-401* to *A-430*.

By the end of 1944 the planes started to show their age after intensive use, so a replacement program was started.
Meanwhile, on 4 January 1945 the Comando en Jefe de Aeronautica (Aeronautic Commander in Chief) formed the Secretaría de Aeronáutica, to which depended the newly created Fuerza Aérea Argentina. All aviation units of the Argentine Army were now centralized. The air force had become an independent force. It was a

time of big changes, as war in Europe was in its closing stages and Argentina, being neutral during almost all of it, couldn't buy new airplanes, so the force now hadn't an effective air power.

By 1 March 1945 the fleet of Northrop 8A was as follows:
Regimiento 3 de Ataque, airplanes in service: *A-401, A-403, A-404, A-406, A-410, A-411, A-412, A-413, A-415, A-416, A-417, A-418, A-419, A-429, A-430, A-420* (on workshops for repairs), *A-421, A-424* (out of service), *A-425, A-427* and *A-428* (transferred to Cordoba).

Plane armed with 11.35mm machine guns, wearing the badge of the Regimiento 3 de Ataque.
(Coll. S. Rivas)

Right:
Emblem of the Regimiento 3 de Ataque on a Northrop 8A.
(Coll. S. Rivas)

While by the end of the year the situation was as follows:

In service: A-403, A-404, A-406, A-410, A-411, A-412, A-415, A-418, A-419, A-425, A-427, A-428 (at Cordoba), A-429 and A-430.

Out of service: A-401 (on repairs after an accident), A-413 (in workshops after an accident), A-416 (main spar being repaired), A-417 (main spar being repaired) and A-420.

The A-407, A-408, A-409 and A-414 were at the Agrupación Entrenamiento in Paraná. Most of the planes out of service were recovered after a major overhaul, but the high number of accidents showed the conditions of the planes, especially regarding the engines, due to a lack of spare parts, as nothing could be purchased between 1941 and 1945. After war ended, spares provision restarted and the operational status was increased considerably.

AFTER WORLD WAR 2

On the provincial elections of 1946 the Northrop 8A received a new task, when the planes A-419 and A-429 were destined, together with the Focke Wulf Fw 44J 156, 181, 182, 186 and 187, to support the elections, under the command of the San Luis Election Command and temporarily based at BAM Coronel Pringles. Later, on 14 April the A-418 crashed at Junín de Los Andes, Neuquén province,

The A-428 in flight. As the original green paint was applied without a primer, the planes started loosing the paint shortly after and looked very bad. (Coll. S. Rivas)

with damage of 70% and with the crew slightly injured. Thirteen days later the most serious accident in the career of the plane occurred when the A-409 crashed near General Pacheco, Buenos Aires province, killing its three occupants. Clearly, the end of the career of the planes on their attack missions was close, as they were obsolete compared to the ones being acquired by other countries of the region.

On 16 June 1946 another accident occurred, when the A-419 was taking off from BAM El Palomar and the pilot noted an engine failure, so he turned to try to return to the runway, but he lost power, crashed and burned, killing its two occupants. The plane was on the base after the military parade held on 4 June for the inauguration of General Juan D. Perón as Argentina's president. On 1 July the A-403 also had an accident while taking off from BAM Coronel Pringles when it was going to Buenos Aires,

Crews preparing for a mission on the mid forties. (Coll. S. Rivas)

Nose of the A-415.
(Coll. S. Rivas)

planes *A-407* and *A-408*, under a request from the Servicio Meteorológico Nacional (National Weather Service), were equipped with systems to take measurements of the atmospheric conditions. These instruments replaced the photo camera and were operated by the observer.

From 1947 onwards, with the arrival of a huge quantity of new planes, which converted the Argentine Air Force into the most powerful in Latin America, different changes took place on the organization of the force. As a replacement of the Northrop at El Plumerillo, FMA I.Ae.24 Calquin attack planes, developed locally, were received from 1947. During 1948 two planes were lost in accidents, the *A-412* near Iguazú Falls in Misiones province and *A-427*

The O-414 when it was used at the Grupo 1 de Observación.
(Coll. S. Rivas)

together with other planes, to take part on the military parade of the Independence Day on 9 July, but the plane was recovered and repaired.

On 30 September a Northrop 8A was deployed to Corrientes, followed by two more on 4 October and the remaining of the Regiment on 10 October to take part on an exercise of the Agrupación Mesopotamia of the Argentine Army. The exercise consisted on close air support to the ground forces, firing with live ammunition, including bombs and machine guns, and also the deployment of smoke curtains. A

total of twelve Northrops took part together with some Junkers K 43 and Fw 44J, operated from field aerodromes and roads, while their crews and mechanics slept on tents. They all returned to BAM El Plumerillo on 15 October.

On 7 December personnel of the Regimiento 3 inspected the runway of Uspallata, Mendoza, for a possible use for training and as a forward field for deployments, in case of war. Shortly after, the idea was left behind.

During September 1947 the Northrop 8A started to perform a new task, when the

at Entre Ríos province, with the loss of all crew members. Also, during that year the *A-410*, *A-415* and *A-418* were retired because they were in very bad condition, two of them damaged beyond repair.

NEW DESTINATION AND ROLE

By Orden Reservada 228 of 2 August 1949, the planes of the Air Force were assigned according to the new organization, which consisted on Brigades to which the groups were attached. The role of the planes switched to that of reconnaissance, so

they received the O of Observation inste-
ad of the A of Attack on their serials. In
the case of the Northrops the order said:
"To the Grupo 1 de Observación at the II
Brigada of Paraná, Entre Ríos Province, are
transferred the following Northrop 8A:
O-401, O-403, O-404, O-406, O-407,
O-408, O-411, O-412, O-413, O-414,
O-416, O-417, O-421, O-425, O-428,
O-429 and O-430 (planes O-407, O-408
and O-414 will be transferred from 15
August". This change was the beginning of
the end for the planes, despite their use
as reconnaissance planes since some time
already. The Northrops replaced an older
plane, the locally built Ae.M.Oe.1 with in-
ferior performance.

During 1949 and 1950 there were no accidents, despite the planes continued flying, accomplishing with the annual plan
of training and participation on exercises,
including some with the Army.

In 1951 there were also no accidents, but
the end of the Northrops was very close
and three planes were retired (O-406,
O-413 and O-425), leaving only thirteen
planes on the unit, which was a big quantity anyway considering the age of the planes. On that year the II Brigada became
the II Brigada Aérea, with the Grupo 1 de
Observación being part of it.
By 19 June 1952, according to Day Or-

Formation of four
planes on their last
years of service.
(Coll. S. Rivas)

Three planes of Gru-
po 1 de Observación
on the late forties or
early fifties.
(Coll. S. Rivas)

der 244, the II Brigada Aérea had eight
Northrop 8A, one de Havilland Dove,
one AT-11 for medical evacuation, one
for photo survey, fifteen FMA I.Ae.DL22
in service and five in storage and 24 FMA

I.Ae.20 El Boyero light planes, but shortly
after this order was modified and the unit
was assigned only with the Northrops
O-404, 416 and 430. Shortly after, again
the assignation of airplanes changed,

The squadron badge of
Grupo 1 de Observación
del Comando Aéreo Ae-
rotáctico

One of the planes on its
last years of service, with
a radio goniometer in-
stalled over the cockpit.
(Coll. S. Rivas)

Escadrille on operations on the late forties or early fifties. (Coll. S. Rivas)

leaving the O-*403, 407, 414, 417* and *418* as added to the unit from 3 September of that year, but they were not operational. Finally, on 22 September 1953 on the Secret Day Order 36 the planes O-404, 416 and 430 are shown as operational on the unit, while the others were already retired. On the last two years of operations there were no accidents, but they rarely flew and on November the last three were retired, being the end of the career of the Northrop 8A in Argentina.

Northrop 8A-2 in Argentina

Serial	c/n	Enlisted	Retired	Notes
401	348	3-6-38	1-1952	Crashed on 7-10-42, 29-10-42 and 13-4-44 at El Plumerillo, and on 29-9-45 and 11-5-48 at Córdoba, repaired. *The 401 after one of its five accidents. (Coll. S. Rivas)*
402	349	3-6-38	25-8-39	Crashed at Punta de Vacas, Mendoza. Pilot Sergeant Victoriano Singer killed.
403	350	3-6-38	1-1953	Crashed on 5-5-44 at El Plumerillo, on 1-7-46 at BAM Coronel Pringles, San Luis, on 2-12-46 repaired.
404	351	3-6-38	11-1953	Crashed on 25-9-42 and 26-2-44 at El Plumerillo, repaired.
405	352	3-6-38	31-10-46	Crashed on 12-12-41. Retired on 3-43. Rebuilt with parts of others in 1946. Crashed on 31-10-46.
406	353	3-6-38	1-1951	Crashed on 23-11-42 at La Paz, Mendoza, and on 11-8-44 at Costa de Araujo, Mendoza, repaired.
407	354	3-6-38	1-1953	Had no accidents during its career
408	355	3-6-38	1-1952	Crashed on 3-9-38, 16-10-40, 4-3-41 and 15-6-44, at El Palomar, the last being a forced landing near Goodyear factory, repaired.
409	356	3-6-38	27-4-46	Crashed on 12-38 and 13-8-44, repaired. Crashed while attemping an emergency landing at General Pacheco. Pilot Cabo Mayor Earle René Laico, observer Alférez Luis Martínez de Alegría and mechanic Suboficial Auxiliar Raúl Garzaro killed.
410	357	3-6-38	27-1-48	Crashed on 10-9-38 and on 6-10-46 at Media Agua, San Juan, not repaired.
411	358	3-6-38	1-1952	Crashed on 30-6-40 and on 21-8-41, repaired. Crashed on 16-3-44 at El Plumerillo, rebuilt in 1945 with parts from other planes. *The 411 after its accident on 16 March 1944 at El Plumerillo Air Base in Mendoza. (Coll. S. Rivas)*

Northrop 8A-2 in Argentina				
Serial	c/n	Enlisted	Retired	Notes
412	359	3-6-38	28-10-48	Crashed in 1939, 25-4-42, 26-9-44 (release of the rear part of the canopy), 13-2-47 (at BAM Coronel Pringles, San Luis), repaired. Dissappeared on the Iguazú River near the falls after hitting the water. Pilot Alférez Jorge Rubén Martínez and mechanic Augusto Díaz killed.
413	360	3-6-38	1-1951	Crashed on 4-3-41, 23-8-43, 1945 and 1946, repaired.
414	361	3-6-38	1-1953	Had no accidents during its career
415	362	3-6-38	27-1-48	Crashed on 3-9-38, 1-7-39, 25-4-47, all at Mendoza, repaired.

The 415 after turning over.

(Coll. S. Rivas)

416	363	3-6-38	11-1953	Crashed on 11-1-43 at Rodeo del Medio, Mendoza and on 1-6-44 (had a mid-air collision with A-419), repaired.
417	364	3-6-38	1-1953	Crashed on 7-10-42, repaired
418	365	3-6-38	27-1-48	Crashed on 4-2-43 at Alsina, Santa Fe, and on 14-4-46 at Junín de los Andes, with damage of 70%, not repaired.
419	366	3-6-38	16-6-46	Crashed on 12-38, 6-7-39, 1943 and 1-6-44 (had a mid-air collision with A-416), repaired. Crashed at El Palomar. Pilot Alférez Eduardo Alberto Sardi and observer Alférez Miguel Angel Bistoletti killed.
420	367	3-6-38	27-1-43	Crashed on 23-11-42, repaired. Crashed on 27-1-43 at El Algarrobal, Mendoza and written off.
421	368	3-6-38	1-1952	Crashed on 4-7-46 at General Zapiola, repaired.
422	369	3-6-38	2-4-41	Crashed at BAM El Plumerillo. Pilot subteniente Ricardo García Roche killed.
423	370	3-6-38	4-3-43	Crashed at Real de la Cruz. Pilot teniente Luis Casimiro Sergio Rodríguez and subteniente Carlos Jensana killed.
424	371	3-6-38	1-7-44	Crashed at BAM El Plumerillo, turned over while taking off and caught fire, being destroyed. Pilot wounded and observer unhurt.

The 424 at BAM Cnel Pringles.

(Coll. S. Rivas)

425	372	3-6-38	1-1951	Had no accidents during its career

426	373	3-6-38	1942	Crashed on 7-9-40

Northrop 8A-2 in Argentina				
Serial	c/n	Enlisted	Retired	Notes
427	374	3-6-38	28-9-48	Crashed in 1938, 9-10-42 at El Plumerillo, 3-12-43 at El Algarrobal, Mendoza, 12-9-46 at El Plumerillo, repaired. Crashed at Colonia Ensayo on 28-9-48, near Paraná, Entre Ríos, when had an engine failure and hit some trees, breaking the fuselage. Pilot Teniente Jorge Alberto Segat unhurt. Passenger Dentist Teniente Natalio Miguel Sangronis killed.
428	375	3-6-38	1-1953	Crashed on 27-5-42 at Villa Nueva, Mendoza, repaired
429	376	3-6-38	1-1952	Crashed on 14-1-44 at El Plumerillo and on 31-8-44 at Campos de los Andes, Mendoza, repaired. *Accident of the 429.* *(Coll. S. Rivas)*
430	377	3-6-38	11-1953	Crashed on 12-3-46 at El Algarrobal, Mendoza, repaired.

Note: In some cases the date of retirement is the one that's on the records and on others is the date of the accident. In some cases, the retirement of a crashed plane was up to two years after its accident.

DOUGLAS 8A-3P IN PERUVIAN SERVICE

By Amaru Tincopa

In the summer of 1937 the *Cuerpo Aeronáutico Peruano* (Peruvian Aeronautic Corps, CAP) launched a modernization program aimed to keep its capabilities up to date with modern air warfare requirements, and to replace a number of aircraft that were becoming obsolescent or approaching to the end of their operative life, such as the Nieuport Delage NiD-121C.1 fighters and the Potez 390A.2 light attack and reconnaissance aircraft, types that eventually found their replacement in the form of the North American NA-50 fighter and Caproni Ca.310 light bomber. This program, however, also contemplated the creation and incorporation into the CAP order of battle of an assault and light attack unit, decision taken after an analysis of the importance of these new aircraft in the Spanish civil war made by the *Comandancia General de Aeronáutica* (Aeronautics High Command, CGA). Along the creation of the assault squadron as part

of CAP organization the CGA ordered the creation of a *Comité Técnico de Adquisiciones* (Technical Committee for Acquisitions, CTA) tasked with the evaluation and utterly selection of the aircraft for use by the new unit.

Nice ¾ view of a Douglas 8A-3P after rolling out of the production line. El Segundo, California, November 1938. (FAP)

The requirements established by the CTA for the CAP assault plane were the following:

- Construction: all metal
- Type: twin seat monoplane
- Engine: radial with an output of no less than 800hp
- Bomb load: not less than 500kg
- Landing gear: retractable
- Armament: 4 x 7.65mm fixed machine guns plus 1 in defensive position.

The CTA started its duties in January 1938 with the evaluation of the Caproni-Bergamaschi AP.1, an aircraft representing the bid offered by *Societá Aeroplani Caproni S.p.A.*, in compliance of the terms contained in the contract signed in 1936 between Peruvian Government and the Italian company.[1] This mixed construction twin seat assault monoplane, however, was quickly discarded by the committee as its technical characteristics did not match the requirements established by CTA. Continuing with the evaluation process, the committee tested the aircraft offered by among others, the following companies:

- *Società Italiana Ernesto Breda per Costruzioni Meccaniche*, with its Ba.65 equipped with type M turret,
- *Industrie Meccaniche Aeronautiche Meridionali* with its Ro.37 equipped with A.30 inline engine,
- Douglas Aircraft Company with the 8A-3 and
- Seversky Aircraft Company with its SEV-2PA-204.

By mid-July 1938 the committee completed the evaluation process and, shortly after, issued his recommendation report to the CGA, in which ruled the Douglas 8A-3 as the winner as this model technical characteristics approached the most to those underlined by the committee. CAP requirements for the 1939-44 periods demanded the purchase of 20 airframes, enough to equip two squadrons, 3 *Escuadrillas*

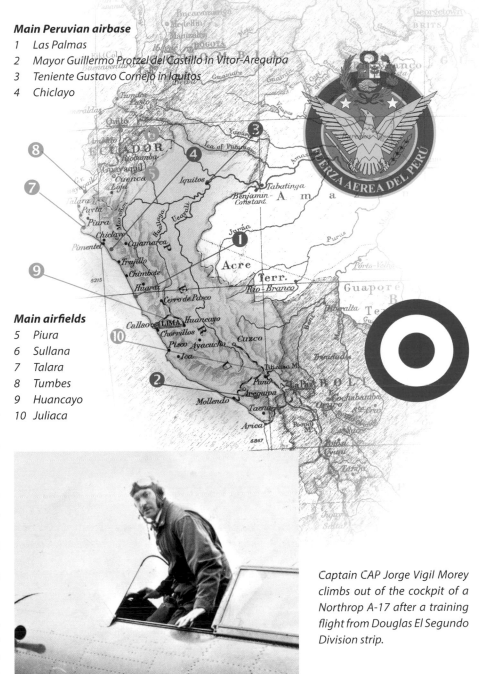

Main Peruvian airbase
1   *Las Palmas*
2   *Mayor Guillermo Protzel del Castillo in Vitor-Arequipa*
3   *Teniente Gustavo Cornejo in Iquitos*
4   *Chiclayo*

Main airfields
5   *Piura*
6   *Sullana*
7   *Talara*
8   *Tumbes*
9   *Huancayo*
10 *Juliaca*

Captain CAP Jorge Vigil Morey climbs out of the cockpit of a Northrop A-17 after a training flight from Douglas El Segundo Division strip.

-or Escadrilles- of three aircraft each, plus one reserve aircraft per squadron. After the CAP requirement was approved by Peruvian Congress and the funds for the purchase were made available, an order for the required number of airframes was issued to the Douglas Aircraft Co., but, before the ink dried on the paper, politics got in the way of the deal as the United States Congress showed opposition to the sale, alleging that the factory was committed to the own US Army Air Corps rearming process. In fact, there were con-

cerns in Washington regarding the sale of military hardware to a government sympathetic with fascism, as the General Oscar Raimundo Benavides administration was considered, and in the end, clearance for the sale of only 10 airframes was given by US government. This decision clearly upset the Peruvians but in the end, there was nothing else they could do about it, and, in mid-August, the contract for the sale of ten model 8A bombers was signed between the Peruvian government and Douglas.

---

1   The contract allowed the opening of an Aeroplani Caproni subsidiary in Lima, and by the terms contained in the agreement the Peruvian government committed to give the Italian company the first option on any future CAP aircraft, unless their characteristics or price were beat by aircraft from other manufacturers.

model reached a total of 550kg, in different combinations, carried both internally on a special bomb bay located behind the pilot, as well as in underbelly racks. These could be launched either by the pilot or by the navigator, depending on the type of attack. The model retained the perforated flaps introduced on the A-17, allowing the aircraft to perform attacks from very step angles.

THE DOUGLAS 8A-3P

Designated as model 8A-3P, the model offered to the CAP was, like its predecessor the 8A-2 offered to Argentina, an export variant of the A-17 "Nomad" assault aircraft, but the Peruvian version differed, in many respects, from its older brother. The main difference was the presence of a retractable landing gear, fitted with big wheels allowing the aircraft operate from even the more primitive airfield conditions. Other improvements included the presence of a bombardier position on a retractable "tub" located on the lower fuselage and the fitting of a more powerful Wright R-1820-G103 "Cyclone" engine, delivering 1000hp on takeoff. These refinements allowed the model to reach a

maximum speed of 383 Km/h at a height of 2,651 meters, and a service ceiling of 7,315m. Internal fuel tanks gave the aircraft an operative range of 1,050km and a combat radius of 490km. Armament was composed by four fixed Browning M1919 .30[1] machine guns in the wings, plus another one mounted on a flexible position operated by the bomb aimer/radio-operator. Also, the bomb load of this

_____
1 Modified to fire 7.65mm ammunition on the Peruvian machines.

On August 17, 1938, *Peruvian Ministerio de Marina y Aviación* (Navy and Aviation Ministry, MMA) issued *Resolución Suprema* (Supreme Resolution, RS) N° 139 ordering the creation of a commission be led by Lieutenant Commander Armando Revoredo Iglesias, and composed by officers Capt. Jorge Vigil Morey, Lieutenants CAP Enrique Espinoza Sánchez and Ernesto Gómez Cornejo, as well as Sub-officers mechanics Oscar Espejo Castro and Federico Vera Principio. These men received the mission to supervise the construction of the Douglas 8A-3P and North American NA-50 airplanes ordered by the CAP. After leaving El Callao harbor in early September the party arrived in Los Angeles a week and a half later, and immediately appointed to the Douglas factory at El Segundo, California, as well as the North American Aircraft Co. plant in Inglewood, California.

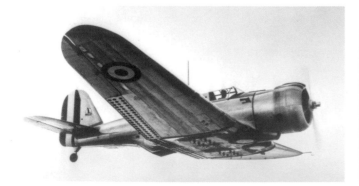

A series of aerial pictures of CAP´s Douglas 8A-3P undergoing test flights. (McPhail)

The beginning of the production suffered from some delays and it was only in early November 1938 when the first 8A-3P rolled out from the production line. The following units followed at a pace of three aircraft per month, and by February 1939 the aircraft order had been completed. Test and evaluation by the CAP commission was completed later that month and after that the aircraft were declared ready for delivery. Only seven of the airframes, however, were unassembled, crated and taken by railroad to Los Angeles harbor, where they were loaded aboard a steamer, leaving for El Callao harbor in late March 1939, while the three remaining airframes remained at El Segundo as Lt.Cmdr. Revoredo had requested the CGA permission to use them in a long distance flight from Los Angeles to Lima. Revoredo, an experienced flyer, wanted to seize the opportunity to put Peruvian aviation on the headlines and, along some of the most experienced officers of the CAP, planned a long range flight leaving from El Segundo, in California, to Limatambo international airport, in Lima, Perú.

After careful preparation and planning, the trio of aircraft[2] left the runway at El Segundo aircraft factory on the morning of May 31, 1939. The aircraft, flown by Lt.Cmdr. Revoredo with Capt. Jorge Vigil Morey as radio operator/flight engineer, Lt. Enrique Espinoza Sánchez, with Oscar Espejo as radio operator, and Ernesto Gómez Cornejo, carrying Federico Vera on

_____

2   Identified with civil serials BO-1G to BO-3G.

Los Angeles, late May 1939. Parked at Douglas El Segundo Division factory tarmac, three of the ten Douglas 8A-3P ordered by Peruvian government in 1938 awaits the beginning of the flight that will take them to Limatambo, Lima. (McPhail)

A nice in flight shot of BO-1G while on enroute to Peruvian territory at the hands of Lt.Cmdr. Armando Revoredo Iglesias. June 4, 1939. (McPhail)

the rear seat. The flight followed the following stages:

1) Los Angeles-El Paso (1158 km)
2) El Paso-Brownsville (1094 km)
3) Brownsville-Tapachula (1234 km)
4) Tapachula-Ciudad de Panamá (1628 km)
5) Ciudad de Panamá-Lima, with a refueling stop at Chiclayo (2453 km)

After covering 7,567 km the flight landed safely in Limatambo international airport in Lima, at 1745 hours of June 5, 1939.

Lt. Espinoza stands in front of XXXI-1 at Limatambo. Note the application of wing chevrons.

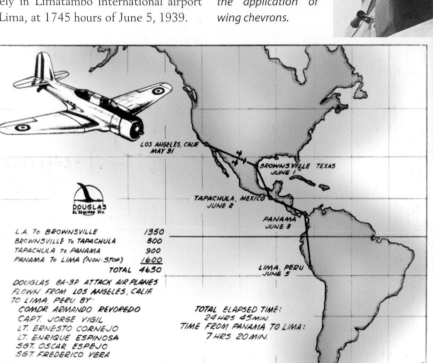

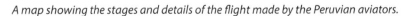

A map showing the stages and details of the flight made by the Peruvian aviators.

IN SERVICE

Once the crates containing the aircraft arrived to El Callao harbor they were taken to "Las Palmas" airfield in Lima where assembly took place by factory personnel sent along the cargo. Once the airframes were erected they were assigned to their unit, the recently activated XXXI Escuadrón de Información y Ataque (XXXI Information and Attack Squadron, XXXI EIA), led by Lt. Cmdr. Armando Revoredo Iglesias. This unit was composed by three escadrilles, the 91, 92 and 93, each equipped with three aircraft, with one aircraft kept as reserve. With the arrival of the remaining aircraft on early June, the unit started an intense operative training cycle at "Las Palmas", its home base, which was completed in September, 1939.

Aircraft from XXXI EIA at "Las Palmas" at the end of 1940 prior to their departure to take part in a navigational training exercise across the country. (Currarino)

Technical data Douglas 8A - 3P	
Engine	Wright GR-1820-G103A Cyclone
Power	1000 hp
Wingspan	47 ft 9 in (14,55 m)
Length	32 ft 1 in (9,91 m)
Height	9 ft 9 in (2,97 m)
Empty weight	4,820 lb (2.186 kg)
All-up weight	7,500 lb (3.402 kg)
Max. speed	238 m/ph (383 km/h) at 8,700 ft (2.650 m)
Range	1180 m (1.900 km)
Service ceiling	24,000 ft (7.315 m)
Climb rate	-
Armament	4 fixed 0.30 in wings, one flexible 0.30 mg
Bombload	20 internally stored 30 lb (13,6 kg) bombs plus four externally carried 100 lb (45 kg) bombs or flares. Maximum bombload 1200 pounds (545 kg).

Seven newly delivered Douglas 8A-3P lined up at Las Palmas airfield in April of 1939.

Below: Members from the Cuerpo Aeronáutico del Perú portrayed at the Douglas El Segundo Division in California shorty before the beginning of the long distance raid from El Segundo-California, to Lima.

Vitor, Arequipa, October 1940. Aircraft from XXXI EIA can be seen next to a pair of Caproni Ca.310 Tipo Perú belonging to the bomber units stationed at "Mayor Guillermo Protzel".
(Maguiña)

Latin American Good Will raid

In the first months of 1940 Revoredo began the planning of an audacious long distance flight around South America, visiting the capitals and major cities of Ecuador, Panamá, Colombia, Venezuela, Brasil, Paraguay, Uruguay, Argentina, Chile and Bolivia in the process, in a feat aimed to demonstrate high level of training and operative capabilities of the CAP. For this mission Revoredo organized an ad-hoc unit composed of five aircraft flown by some of the best CAP airmen, such as Captains Enrique Bernales Bedoya and Jorge Vigil Morey; Lieutenants Ernesto Gomez Cornejo, Manuel Gambetta Pielago, Pedro Vargas Prada, Cesar Cossio Tudela and Jesus Melgar Escutti, along with Sub Officer Mechanics Federico Vera and Ernesto Green. Revoredo christened these men as Los Zorros (The Foxes).

The raid, scheduled to be completed in 15 stages, began on this unit left Limatambo international airport on the evening of March 23 heading to the city of Quito, Ecuador, as their first stop, arriving to the "Mariscal Sucre" international airport the next morning. As the unit was preparing to leave the Ecuadorian capital bound for the city of Bogotá, Colombia, an incident took place as Capt. José Bernales, aboard Douglas serial XXXI-2, was taxiing on its way to perform takeoff when he got –apparently- distracted by cause of some local females waving the men from the edge of the runway, hitting a boulder located at the edge of the runway. As consequence of the impact one of the wingtips suffered moderate damage and the aircraft was rendered unfit for flight. Revoredo ordered Bernales and a mechanic to await the arrival of a spare wing from Lima while he and the remaining of the group continued to Bogotá. The wing – which

Quito, March 23, 1940. Several views of the men and aircraft from "Los Zorros" escadrille shortly after their arrival to "Mariscal Sucre" airport. As part of their Latin-American goodwill flight. (IEHAP)

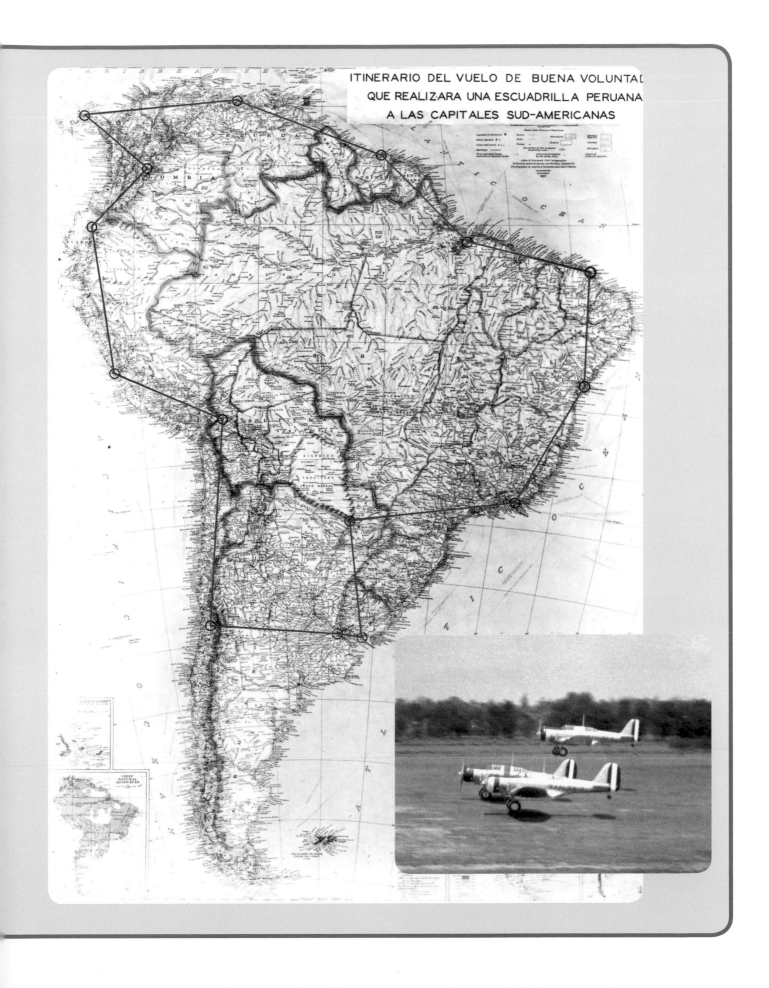

ITINERARIO DEL VUELO DE BUENA VOLUNTAD
QUE REALIZARA UNA ESCUADRILLA PERUANA
A LAS CAPITALES SUD-AMERICANAS

Picture showing the four remaining aircraft from the flight during their visit to the Brazilian city of Rio de Janeiro.
(IEHAP)

arrived to Quito tied to the underbelly of a CAP Curtiss BT-32 "Condor"! - was quickly mated onto the XXXI-2 airframe and both aircraft headed back to Lima. [Meanwhile, Revoredo had arrived to "El Techo" airport at the Colombian capital, and from there – and after carrying the required protocol activities- continued their journey without incident, visiting the cities of Caracas-Venezuela, Paramaribo, Belen do Pará, Fortaleza, Rio de Janeiro-in Brazil, Asunción-Paraguay, Montevideo-Uruguay, Buenos Aires-Argentina, Santiago de Chile and La Paz, in Bolivia before returning to Lima on May 3, after completing a total flight distance of over 17,900km.

Aircraft from the "Los Zorros" escadrille seen at Limatambo international airport shortly after their return to Lima.
(IEHAP)

On April 12, 1941, almost a year after completing the famous raid and while taking part in a nationwide event aimed to raise funds for the "national armed forces reserve", tragedy struck XXXI EIA when Lieutenants Luis Cossio Tudela and

José Voto Bernales were killed after the Douglas 8A-3P they were flying fell to the earth near Ckari, located on the outskirts of Cusco, in the middle of the Andes, due to an engine failure.

After the ceremony held at Limatambo Peruvian president Manuel Prado Ugarteche was invited by Lt. Cmdr. Armando Revoredo to overfly the city.
(De la Puente)

A pair of beautiful shots from XXXI EIA aircraft at Arequipa airfield, with the "Misti" volcano serving as background. October 1940. *(Currarino)*;

The squadron also paid a visit to Juliaca forward airfield as part of its navigational training. *(IEHAP)*

Right: XXXI EIA also visited Huaraz airfield, in Ancash, as part of its navigation training. Here their Douglas can be seen along Caproni Ca.111 bombers and North American NA-50 fighters from XI Escuadrón de Bombardeo and XXI Escuadrón de Caza, respectively. *(DIRAF)*

93 Escuadrilla takes off from "Las Palmas" during a training mission in early 1941.
(IEHAP)

IN ACTION OVER ECUADOR

During the first days of the month of July, 1941, the poor diplomatic relationships situation between Ecuadorian and Peruvian governments, which had been markedly deteriorating since late 1939, broke into a fully armed conflict on July 5. Ecuadorean army elements attacked Peruvian posts along the border, actions that provided the Peruvian government with the perfect excuse to begin a large military operation against Ecuadorian military forces along the border, aiming to destroy them and force the government at Quito to the negotiations table thus ending, once and for all, the border dispute.

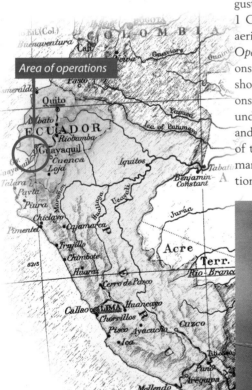

Area of operations

The XXXI EIA, while not taking part in the operations from the beginning, was deployed to the area of operations on August 24 and put under the command of 1 Grupo Aéreo (1 Air Group), cluster of aerial units subordinated to the *Teatro de Operaciones del Norte* (Northern Operations Theater, TON[1]). While the XXXI EIA should have been involved in the operations from the very beginning, lack of vision, understanding of the modern air warfare and the appropriate knowledge of the use of tactical aircraft by the TON high command kept this unit excluded from the action until CAP losses began to mount.

That very same morning the XXXI EIA began its deployment by leaving "Las Palmas" AB at 1000hrs heading north bound to *"Teniente Coronel Pedro Ruiz Gallo"* airbase in Chiclayo, where the unit arrived two hours and a half later. Shortly after, at 1400hrs, the three ships from 91 escadrille left Chiclayo heading for "Capitán Guillermo Concha Iberico" secondary airfield in Piura. Aboard the aircraft were the Aeronautics' Chief of Staff and the head of TON, who spent the night on that location. The following morning, the 1 GA issued a directive for the 91 Esc. ordering to establish its command at Tumbes for-

---

1   An Ad-Hoc arrangement of aerial, ground and naval forces created in January 1941 with the purpose of protecting the Peruvian northern territories.

Aircraft from XXXI EIA seen during their deployment to the north of the country in order to take part in the operations against Ecuadorian forces. July 24, 1941. (Currarino)

Aircraft from XXXI EIA seen during their deployment to the north of the country in order to take part in the operations against Ecuadorian forces. July 24, 1941. (Currarino)

ward airfield, an order that was accomplished by 1130hrs. A day later, on the 26th, 1 GA Command ordered the transfer of the unit´s two remaining escadrilles from Chiclayo to Talara, where they were stationed awaiting further orders.

With all its units deployed in the operations zone, the XXXI EIA finally entered the fray on July 27 when the 91 Esc. received an order to perform a reconnaissance sortie over the towns of Cuenca and Santa Rosa, located inside Ecuadorian territory. This mission had the primary objective to discover any Ecuadorian army troop movement around these towns as well as to assert the condition of the railroad bridge at the town of Uzhcurrumi, previously attacked by Ca.310 bombers from XI Escuadrón de Bombardeo (XI Bombardment Squadron, XI EB), while it's secondary objective included the reconnaissance over the Girón area. The mission was launched at 1645hrs when aircraft *XXXI-91-1* and *XXXI-91-3* took off from Tumbes heading north, following the Puerto Bolívar-Machala-San Sebastian-Santa Isabel-Asunción-Girón and Tarqui itinerary, and accomplishing the primary mission. Bad weather, however, prevented the reconnaissance over Uzhcurrumi Bridge and the aircraft returned to their base, landing at Tumbes without incidents at 1812 hours.

Aircraft from XXXI EIA at Talara airfield shortly before their departure to Tumbes forward airfield.

(Currarino)

On July 29 the XXXI EIA received an order from the chief of Primera División Ligera (First Light Division, 1 DL) requesting, as primary objective, the destruction of the bridges at Federico Páez and Uzhcurrumi, as well as the attack of the bridge at Arenillas as well as Ecuadorian Army units stationed around Arenillas and Chacras along any opportunity targets located in the area. The aircraft, armed with 6 x 13.5kg anti personnel and 4 x 50kg demolition bombs each, left Tumbes at 1115hrs, flying to their intended objectives and attacking the bridges with the 50kg devices, destroying them. Later, EA troop concentrations were discovered and duly attacked with machine gun fire and anti-personnel bombs. Finally, after completing their mission, the escadrille returned to Tumbes airfield at 1330 hours. The next day would mark the operational debut in the operations area of 92 Escadrille after it received an order to attack Ecuadorian defensive positions discovered on the riffs

and hills around Arenillas, as well as in the right bank of Santa Rosa River, with the secondary mission to perform a photographic reconnaissance of the road connecting Arenillas and Chacras, the Santa Rosa´s railroad station and the highway connecting this city with Arenillas. To accomplish this order Douglas *XXXI-92-2* and *XXXI-92-3*, each armed with 60 2kg anti-personnel sub-munitions, scrambled at 1045hrs heading to their designed targets and attacking troop concentrations discovered in the right bank of the Santa Rosa river and around Arenillas, with great success. After completing this first sortie the aircraft returned to Talara, where their crews received the order to rearm their aircraft and return to the area to repeat the attack, receiving the secondary mission to perform a reconnaissance following the Chacras-Arenillas-Santa Rosa-Pitahaya-Puerto Bolívar itinerary. After successfully completing their assigned missions, crews landed in Talara safely at 1520hrs repor-

Beautiful view of a pair of Douglas 8A-3P from 92 Escuadrilla flying over the typical landscape of Peruvian northern coastal territories. July 1941.
(IEHAP)

ting during their debriefing that there was a complete absence of Ecuadorian forces movement in Arenillas, Santa Rosa and Puerto Bolívar, suggesting that the blunt of EA units had withdrawn from these areas. Meanwhile the 91 Esc. was also busy performing a reconnaissance flight over the Quebrada Seca-Jubones-Girón-Tarqui-Cuenca area looking for any EA movement around these areas.

On the morning of July 31 the 1 GA Command issued several orders for the XXXI EIA escadrilles, which translated in a number of sorties launched in support of the airborne operations scheduled to take place that day, which ended with the capture of Machala, Santa Rosa and the Puerto Bolívar. The specific orders for the XXXI EIA units included:
- 91 Escadrille: attack EA troop concentrations located around the town of

Pasaje, as well as to provide air cover to the Peruvian army forces advancing over Puerto Bolívar, Pasaje, Machala, Santa Rosa, as well as close air support for the airborne forces tasked with the capture of Santa Rosa, Machala and Puerto Bolivar.
- 93 Escadrille: shift its operations center from Talara to Tumbes.

Operations began at 11.15 hours with the transfer of 93 Esc. to Tumbes, where its

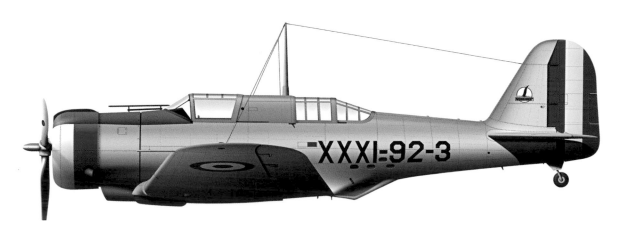

A 92 escuadrilla machine as it appeared during the conflict with Ecuador. (Profile by Alexandre Guedes)

commander received the order to attack on Ecuadorian positions around Macará, with the secondary mission to perform a reconnaissance over the surroundings. The first objective of the mission was completed without incidents, with the crews reporting that EA units had fled the area in front of the Peruvian army advances. However, while flying between La Toma and Macará the unit was attacked by anti-aircraft fire, prompting the reaction of the Peruvian aviators who in turn strafed the Ecuadorian positions. After landing in Tumbes, close inspection revealed several bullet impacts on its fuselage of Douglas s/n *XXXI-93-3*, evidencing the intensity of the combat.

91 Esc. aircraft also performed an armed recognition sortie over Puerto Bolívar, during which a number of barges employed to discharge provisions were discovered and strafed before the unit proceeded eastwards towards Machala, where an horse drawn convoy, loaded with supplies, was attacked. Between the Girón and Tarqui, an EA truck convoy was sighted and attacked employing 13.5kg anti-personnel bombs, causing heavy losses among Ecuadorian forces. Just 30km from there, on the route between Girón and Pasaje, the flight discovered two infantry companies which tried, unsuccessfully, to camouflage in the surroundings, and attack them. Finally, while on their way back to Tumbes, the aircraft strafed several barges discovered on the tidelands that surround Puerto Bolívar.

The first day of August, and with the cease of fire agreement already in full effect, the XXXI EIA received orders to send an escadrille over Puerto Bolivar and Machala with the mission to cover the landing of Ejército Peruano (Peruvian Army troops, EP) troops on these locations. This mission was trusted to the 91 Escadrille and accomplished without incidents. On August 5, TON Command requested a reconnaissance flight over the towns of Pasaje and El Guabo as well as over Uzhcurrumi Bridge in order to inform about any possible presence of any EA elements on these areas. The mission was completed by the men from 92 Escadrille which reported on

Ground personnel rearm this Douglas with fragmentation bombs prior to its departure in yet another mission. Tumbes forward airfield, late July 1941. (IEHAP)

Mechanics work on the Wright R-1820 engine from this 8A-3P at Tumbes forward airfield, late July 1941. (DIRAF)

their debriefing that local population had returned to their home towns. Simultaneously, 1 GA command ordered the XXXI EIA to transfer the 93 Escadrille from Talara to Tumbes. The next morning the 91 Esc. deployed Douglas s/n *XXXI-91-2* and s/n *XXXI-91-3* on a escort mission in behalf of 105 Escuadrilla de Transporte (105 Transport Escadrille, 105 ET) which sent a Caproni Ca.111, identified with serial 105-5, carrying a high ranking EP officer

to that location. A reconnaissance over the surrounding of Machala was assigned as secondary objective, missions accomplished without incidents

August 8 saw no activity for the XXXI EIA and its personnel was allowed to rest while aircraft was provided with maintenance and that afternoon orders arrived for the Douglas s/n *XXXI-93-3*, damaged by AA fire a few days before, to be flown back to

Lieutenants Lynch, Estremadoyro and Leon (from right to left) stand in front a lineup of XXXI EIA aircraft. While the two first served with XXXI EIA, the later was an officer assigned to the XI Escuadrón de Bombardeo.

cillo-Carlomagno-Sabiango-Macará areas. This mission was launched at 1350 hours and was accomplished without incidents.

On August 22, Douglas s/n *XXXI-93-3* returned to its unit after completing its repair works at Las Palmas. Two days later, the United States military attaché, Colonel Urzal Ent, was flown from Tumbes to Santa Rosa aboard Douglas s/n *XXXI-91-1*. The following day the TON HQ ordered the XXXI EIA to perform of an armed recon patrol over the Pasaje-Girón axis, the road towards Pucará, Guanazan and Yulo in order to discover any enemy forces in these areas, as well as their number, direction and. This mission was trusted to the 92 Esc. which deployed Douglas *XXXI-92-1* and *XXXI-92-3* from Tumbes at 1235 hours. Upon return, almost two hours later, the crews reported the presence of Peruvian troops around the town of Pasaje, which were almost attacked by mistake.

On the morning of September 9 the command from XXXI EIA ordered the transfer of Douglas s/n *XXXI-93-1* to Las Palmas in order to receive repairs and major maintenance. After some days absent from operations, the XXXI EIA returned to action on the evening of September 11 when TON HQ ordered the air attacks to be held against any Ecuadorian Army units found near the Porotillo sector, in retaliation for a bloody ambush which took place at 1300hrs against an EP platoon patrolling the area and caused heavy losses.

Las Palmas to be repaired. August 9, however, saw the Squadron back in action when hostilities renewed due to the violation of the cease fire agreement by part of Ecuadorian forces. TON High Command ordered its aerial units to perform reconnaissance, and later, retaliation strikes against Ecuadorian military targets located in Guayaquil and Cuenca, including the railway stations in of the before mentioned cities. The following day the 1 GA command ordered the squadron to deploy one of its escadrilles to Sullana forward airfield, located in Piura, and during the morning the 91 Esc. was flown to that airfield. From

this location, the escadrille flew an armed reconnaissance sortie that covered the Macará-Zabiango-Carlomagno-Loja-Célica-Laceiba-Zapotillo-Alamor area, before returning to Sullana. Meanwhile, 92 Esc. was withdrawn to "Capitan Guillermo Concha Iberico" forward airbase near Piura in order to rearm as the bomb stock at Tumbes has been depleted. After arming each aircraft with four 13.5kg bombs, the unit took off and headed to Sullana, where it received orders from Octava Division Ligera (8 Light Division, 8 DL) Chief of Staff, which requested an armed reconnaissance patrol over the Alamor-Zapotillo-La Ceiba-Sau-

Douglas attack planes and Caproni transport can be seen resting between sorties in Tumbes forward airfield. August, 1941. (IEHAP)

On September 14 the Command of XXXI EIA ordered all its units to perform attacks on EA troop concentrations located in and around Pucara, Santa Isabel, Jubones canyon and Uzhcurrumi. 93 Esc. was the first into action, with Douglas s/n *XXXI-93-1* and *XXXI-93-3*, each armed with 40 x 2kg antipersonnel sub-munitions, departing from Sullana at 10.45 hours and heading north east towards their intended targets. After reconnoitering over the Pasaje-Porotillo-Uzhcurrumi-Girón region and while overflying Porotillo, the flight noticed enemy troop movement below, proceeding to attack immediately with machine gun fire and antipersonnel bombs. Finally, near the town of Sharo, the flight also discovered nearly 15 campaign tents, which were duly strafed before returning to base. Meanwhile, the 92 Esc. also took part on operations that day by launching a two ship armed reconnaissance patrol over Santa Rosa. This mission involved Douglas s/n *XXXI-92-2* and *XXXI-92-3* which left at 1525hrs, arriving to its target about twenty minutes later only to found bad weather over the area, preventing them to perform their primary mission. During their return and while overflying Pucará, however, the formation sighted a mule-drawn convoy leaded by several dozen EA soldiers and proceeded to attack them with machine gun fire, causing heavy losses. Later, and while overflying Santa Isabel, the flight identified a military camp, which was also strafed and bombed. Low on fuel and ammunition, the pair of aircraft returned to Sullana at 1630 hours.

At 0900 hours from September 15, 1941, the 1 GA ordered a two ship formation from 91 Esc. to perform an attack on Ecuadorian army positions around the Guapo-Jubones river mouth-Dos Bocas and Tendales line. The crews assigned to accomplish the mission took off aboard Douglas s/n *XXXI-93-1* and *XXXI-93-2* and, once over the target area, they discovered EA personnel performing supply duties on the docks near Jubones river mouth, attacking them. Over Dos Bocas the flight also found a number of EA mule-drawn carts, which were also subject of attack. It is worth notice that while reconnoitering over the Jubones River, the formation was

attacked by light and medium caliber AA fire, and the Peruvian aircraft returned the fire, managing to silence the positions but sustaining some impacts in the process. Continuing with the assigned missions the flight headed towards the towns of Barbones Bajo and Barbones, where it attacked a group of Ecuadorian soldiers. Meanwhile, 92 Esc. launched a pair of aircraft, Douglas s/n *XXXI-92-1* and *XXXI-92-3*, at 1100hrs on an armed reconnaissance sortie over Tendales where they reported the presence of EA troop movement. These, alerted by the sighting of the Peruvian aircraft, the soldiers quickly seek cover inside a building, but it was quickly turn into rubble after the Peruvian aircraft dropped a pair of 50kg bombs on it. Next, the patrol headed to Tenguel where their pilots discovered a pair of abandoned AA gun emplacements, destroying them. Finally, the flight discovered a number of minor vessels located north of La Puntilla, as well as EA troops on the nearby docks, which were attacked until the aircraft ran out of bombs and ammunition. Both airships then returned to Sullana at 1245hrs, where the ground crews reported the use of 1,184 FN SS-34 7.65mm cartridges, 20 x 13.5kg fragmentation bombs as well as two x 50kg demo-

lition bombs. Meanwhile, at 1130hrs, the 91 Esc. executed an order issued by TON command, which had requested a reconnaissance sortie to be flown over Chacras, Arenillas, Santa Rosa, Santo Tomas, Pasaje, San Francisco, Jubones mouth and El Guabillo. Upon debriefing the crews reported that the Ecuadorian forces had completed the evacuation of Pagua and Tendales, as well as the attack on retreating EA forces discovered near Tenguel. Also, in Balao-Embarcadero the flight discovered seven transport barges, of approximately 30 feet each, along eight barges for livestock, which were attacked in concert with a Curtiss model 37F "Cyclone Falcon" from 82 Esc., XXXII Escuadrón de Información Marítima. Finally, while heading back to base the flight discovered three sailing ships, of about 30 to 40ton each, and proceeded to attack them with machine gun fire, sinking all of them. Total ammunition expended during this mission accounted for a total of 3,528 FN SS-34 cartridges and 84 x 2kg sub-munitions.

On September 19, 1 GA Command issued two orders for the XXXI EIA. The first requested an armed reconnaissance sortie over the Piedras-Moromoro-Portovello-

Ground personnel covers from the inclement sun under the umbrella provided by this Douglas 8A-3P from XXXI EIA at Talara forward airfield, 1941. A pair of Ca.310 bombers and at least a Ca.114 fighter can be also seen parked at this airfield, which served as forward repair center for aircraft deployed to the TON.
(De la Puente)

CAP personnel inspect the damage sustained by this Douglas after a landing mishap at Villa auxiliary airfield. October, 1941.

Zaruma axis, while the second called for the attack on any Ecuadorian forces discovered in the before mentioned areas. First mission was assigned to the 93 Esc. which launched a pair of Douglas, s/n *XXXI-93-2* and *XXXI-93-3*, from Sullana at 1345hrs. The flight headed to the town of Piedras, where their crews discovered an EA cavalry section, which was strafed. The pair then continued towards its secondary targets before returning to its base without incidents. The second mission was assigned to 91 Esc. which deployed Douglas s/n *XXXI-91-1* and *XXXI-91-2* at 1700hrs with the primary objective to perform an armed reconnaissance sortie over Piedras-Panupalis-Sucaray-Portovello-Piñas-Moromoro looking for targets of opportunity. A friendly fire incident took place when EP troops were attacked by mistake when the pilots mistook them for Ecuadorian forces about 5km east of Panupalis, fortunately without causing any losses. During the return flight and while overflying the town of Piñas, the pilots discovered an EA artillery placement which was duly attacked with bombs. During this mission a total of 383 FN SS-34 cartridges as well as 20 x 13.5kg

and 10 x 11kg fragmentation bombs were expended.

On October 2 the hostilities came to an end thanks to the signature of a peace agreement, known as the Acuerdo de Talara, by the opposing sides. This was the first step towards a definitive peace agreement signed in Brazil, known as the Protocolo de Paz, Amistad y Límites de Río de Janeiro, signed on January 29, 1942 in Brazil. By the terms of the agreement, both sides began the demilitarization of the border and, consequently, the TON HQ ordered the gradual withdrawal of all its composing units either to rear areas or, in some cases, directly to their home bases. XXXI EIA began its withdrawal on that very same date, when it left Sullana for Talara at 1310hrs, arriving shortly after at its destination. From Talara the unit carried out the following exercises:

October 2: Blind formation flying.
October 3: Formation flying with XXI Escuadrón de Caza units
October 4: Level and dive bombing training.
October 5: Navigational training along XI EB and XXI EC units.

On October 5, a Caproni Ca.114 from 42 Esc., s/n *XXI-42-2*, and a Douglas from 91 Esc., s/n *XXXI-91-1*, were deployed to Tumbes, with the remaining elements from XXXI EIA the next morning. On October 15, the TON HQ issued movement directive Nº1 to the commands from XI, XXI and XXXI Squadrons, ordering the XXXI EIA to deploy one of its units to Tumbes forward airfield and another to Talara. The remaining unit was ordered to return to Las Palmas for repair and rest.

In early November the CGA ordered the return of all units assigned under the command of TON to their home bases, order which included the two XXXI EIA escadrilles still deployed to the north of the country which completed their return to "Las Palmas" on the evening of November 6, 1941. A leave was granted for flight and ground crews while all aircraft were sent to the Arsenal de Aeronáutica for a mandatory overhaul on their structures after three months of constant operations from the poorly prepared airfields found in the conflict area.

The year 1941 did not end without tragedy as, on October 16, while carrying out a training flight near the Escuela de Vuelo de Altura (High Altitude Flight School) located in Huancayo airfield, in the middle of the Andes mountain range, 2nd Lt. Cesar Benavides Bielich and 3rd class Sub Officer Tito Martin Lynch were killed when their aircraft crashed against a hill near the Huaynacocha lagoon in Junín, due to bad weather.

BACK TO "LAS PALMAS":
OVERHAUL AND NEW ASSIGNMENTS

The signature by Peruvian and US governments of the "Military Assistance Program" (MAP) in late 1941, followed by the Lend Lease agreement on March 11, 1942, and, shortly after, of the Hemisp-

By April 1942 the squadron was back to full strength after the last aircraft left the *Arsenal de Aeronáutica*[1] after completing their overhaul process. In order to effectively comply with their assigned task, the aircraft received a paint finish consisting of an overall Non-Specular Sea Blue paint finish, which blended very well with the Pacific Ocean. For identification purposes the leading aircraft of each escadrille received red and blue paint on their cowlings. It is worth mentioning that the aircraft also received a life raft fitted on the rear canopy in order to increase the survival chances of the crew in case of ditching.

---

1   Name given to the Caproni Peruana aircraft factory after its nationalization by Peruvian government in 1941.

ENTERS THE DOUGLAS 8A-5N

The CGA in Lima became aware of the availability of the surplus ex-Norwegian Douglas 8A-5N aircraft since early 1941, and sent a request to purchase these available planes in order to complete the XXXI EIA aircraft strength as established in the original strategic planning for the years 1939-1944, and therefore negotiations began in Washington after the Peruvians notified the US government about their interest in the purchase of the ex-Norwegian aircraft. The US Defense Department, however, opposed the sale as they feared that the aircraft could be used against Ecuador. Peruvians insisted again in August 1941, only to be rejected again. Finally, in June 1943, the US Congress agreed the transfer

Peruvian president Manuel Prado Ugarteche (standing on the wing next to the cockpit) receives details about the characteristics of the Douglas 8A-5 by part of a CAP officer shortly after the arrival of these machines to Lima. Of interest are the provisional serials painted on the aircraft, as well as the "Misti" name, complete with the silhouette of the famous volcano, applied on the cowling. (IEHAP)

Below: *Although of poor quality, this rare picture shows the dark blue overall scheme adopted by the Douglas 8A-3P after the disbandment of XXXI EIA and following assignment to the newly activated 23 EIA. (IEHAP)*

heric Defense Cooperation Agreement between Peruvian and US governments on April 24 of the same year, opened a new chapter for the Peruvian armed forces with its commitment to the defense of the Allied forces strategic assets inside Peruvian territory, such as the oil refineries in Talara and the raw material shipments from Callao and Chimbote harbors against possible attack by axis forces. XXXI EIA became an important part on the CAP strategic scheme for the Peruvian coastal defense, receiving the task to patrol the central and northern coastal areas of the country, searching for any suspected activity from Japanese warships, in special submarines.

Right: Pilot instrument panel detail

The electrical and engine commands from a 8A-5N as seen on pilot´s port side position.

Detail view from the radio installed on the 8A-5N operated by the Peruvian Air Corps.

Left: Pilot cockpit panel, port view.

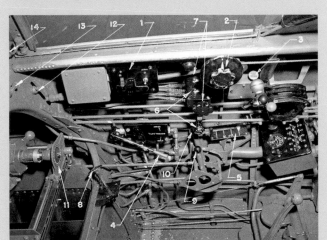

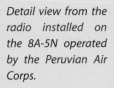

.Starboard panel view from the 8A-5N pilot cockpit.

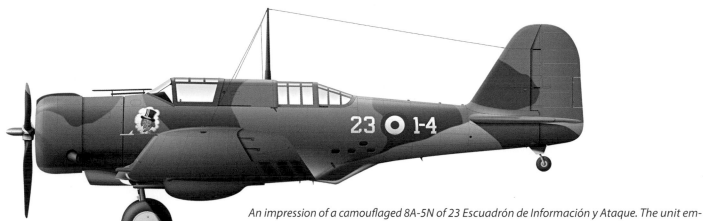

An impression of a camouflaged 8A-5N of 23 Escuadrón de Información y Ataque. The unit emblem was created around a stylized number 23

(Profile by Alexandre Guedes)

of the remaining 13 airframes,[2] along two sets of propellers and five engines[3], to the CAP under the auspice of the MAP.

After years of use and storage in less than ideal conditions, at the time of the sale the surviving aircraft were in poor condition, and therefore, and as part of the agreement, the US government committed to perform a major repair, including the fitting of brand new engines, to all of the airframes before their transfer to the CAP. In the meantime, Peruvian *Ministerio de Aeronáutica*[4] (Aeronautics Ministry, MA) designed a committee composed by several CAP officers to travel to the US in order to supervise the overhaul and acceptance process for the procured airframes. This party was followed, in late October 1943, by a number of officer pilots from the newly activated *23 Escuadrón de Información y Ataque* (23 Information and Attack Squadron, 23 EIA) and specialists who were tasked with the ferry of the 13 airframes from US to Peruvian territory. The aircraft completed their overhaul receiving a Dark Green overall paint scheme, and were ready for acceptance and transfer to Peruvian territory the first days of November, 1943.

---

2   It is worth notice that an 14th airframe was eventually sent to Perú as attrition replacement.

3   This airframe was eventually fully repaired at the Arsenal de Aeronáutica and pressed into service.

4   Secretary created on October 27, 1941, replacing the CGA.

A nice view of a 8A-5N from the Peruvian Aeronautic Corps taken at "Las Palmas" in 1944. Note the unit emblem and the .50 machine gun gondolas under the wings.

Due to the risks imposed by German and Japanese submarines to shipping, it was decided that the Peruvian 8A-5N[5] should be ferried to Lima along ten Curtiss Hawk 75A-8,[6] and, after careful planning and preparation, on November 16, 1943, the Douglas and Curtiss left Kelly field heading towards Brownsville, from where they headed south following the following stages: Veracruz-Tapachula (México) - Managua (Nicaragua) - Camp David – Rio Hato (Panamá) - Cali (Colombia)-Talara and Lima (Peru).

The flight, lead by Lt. Cmdr. Luis Cayo Murillo, was organized as follows:

[5] The aircraft received new USAAC serials, only for record-keeping purposes, running from 42-109007 to 42-109019.

[6] Part of a 28 aircraft total purchased under the Lend Lease Program in 1942.

1st Squadron

8A-5N registry N° 23-C, Command aircraft:		Lt.Cmdr. Luís Cayo Murillo S.O. Especialist José Hernández
Leader 1st flight (8A-5N)		
Douglas	23-1-1	Lt.Cmdr. Enrique Bernales B. S.O. Especialist Victor Raffo.
Douglas	23-1-2:	Capt. Enrique Escribens. S.O. Rubén Espejo.
Douglas	23-1-3	Lt. Alberto López C.
Douglas	23-1-4	2nd Lt. Hernán Souza P. 2nd Lt. Cesar Benavides G.
Leader 2nd flight (Hawk 75A-8)		
Curtiss	21-1-1	Cap. César Bielich M.
Curtiss	21-1-2	Lt. César Garcés C.
Curtiss	21-1-3	2nd Lt. Rolando Gilardi R.
Curtiss	21-1-4	2nd Lt. Hector Alfaro R.

2nd Squadron

Douglas 8A-5 serial 23-2-1, flight leader		Lt.Cmdr. Guillermo Van O`ordt. S.O. Luis Veraud.
1st flight (8A-5)		
Douglas	23-2-2	Capt. Jaime Cayo Murillo S.O. Ernesto Green T.
Douglas	23-2-3	2nd Lt. César Podestá J.
Douglas	23-2-4	2nd Lt. Eduardo Santa María R.
2nd flight (Hawk 75A-8)		
	21-2-1	Capt. Alberto Botto E.
	21-2-3	Lt. Gustavo Salcedo S.
	21-2-4	2nd Lt. Jorge Chamot B.

3nd Squadron

Douglas 8A-5N serial 23-3-1, flight leader		Capt. Jesús Melgar E. S.O. Roberto Reyes.
1st flight (8A-5)		
	23-3-2	Capt. Ricardo León de la Fuente
	23-3-3	Lt. Ernesto Fernández L.
	23-3-4	2nd Lt. Pedro Melgar E. 2nd Lt. Luis Robles A.
2nd flight (Hawk 75A-8)		
	21-3-1	Capt. César Lynch Cordero.
	21-3-2	Lt. Alfonso Terán B.
	21-3-3	Lt. Roberto Wangerman C.

After completing the long journey the formation arrived in Lima without incidents on November 19, completing a total flight time of 29 hours, 45 minutes, facing all kind of atmospheric conditions.

An additional 8A-5N airframe was delivered to the CAP in late 1944 after it completed its repair in US territory. This aircraft arrived to the country still sporting the Norwegian colors, as can be seen in this picture from December 1944. Note that this particular airframe was devoid of underwing gondolas.
(H. Currarino)

8A-5N OPERATIONS WITH THE 23 ESCUADRÓN DE INFORMACIÓN Y ATAQUE

After the official reception, the 13 8A-5N bombers were deployed to "Capitán Victor Montes" airfield in Talara, Piura, where it would serve during the following two years. Meanwhile, the MA ordered the disbandment of the 31 EIA, transferring their remaining aircraft to Talara in order to increase the strength of 23 EIA.

Always from Talara, the new unit began to perform patrol duties along the northern Peruvian coastline in cooperation with those from the USAAF stationed on the nearby "El Pato" airbase,[1] located a few miles east of Talara.

In early 1944 the 23 EIA aircraft were sent to the Arsenal de Aeronáutica for inspection, receiving a new Dark Green, Dark Earth and Light Grey paint scheme.

An additional 8A-5N was delivered to the Peruvians in November of 1944.
This particular aircraft was delivered in full Royal Norwegian AF markings and was devoid of the characteristic underwing gondolas.
(H. Currarino)

With the end of the war came the end of the of economic aid given by the US government to its "backyard allies" armed forces, including Perú, which has been flowing steadily under the auspice of the Defense Agreements from 1941 to 1945. This situation forced the Peruvian MA to introduce a number of modifications to its structure and organization, reducing the number of operative unit to realistic numbers. This measure affected the 23 EIA which, along other units, became disbanded in January 1946. Simultaneously, the MA reactivated the old 31 EIA, operating a mixture of 8A-3P and 8A-5N from "Las

Palmas" AB in Lima, with Collique AB, in the outskirts of Lima, as secondary airfield. Other important changes included updates to the aircraft registry system, with a new registration scheme divided in seven organized numeric blocks, as follows:

100 fighters
200 fighter bombers and assault aircraft
300 transports
400 training
500 miscellaneous
600 helicopters
700 liaison

---

1   Airbase built by the US government in accordance to the defense agreement signed between the US and Peruvian governments on April 24, 1942.

A delegation of CAP and US Military mission officers, headed by president Prado, review the men and machines deployed to "Capitan Victor Montes" airbase in Talara on June 26, 1944.

Douglas from the reformed 31 Escuadron de Informacion y Ataque prepare to take off from the runway at "Teniente Coronel Pedro Ruiz Gallo" airbase, Chiclayo, in July 1947.

Escuela de Aviación Militar student cadets pose next to a Douglas 8A-5 from 23 Escuadrón de Información y Ataque at Las Palmas airbase, August 1944. (Currarino)

By this time, the squadron aircraft had reverted to a natural metal finish, wearing only a black anti-glare on the upper nose as well as a color band, with the letters A, B or C in the middle, in the rear fuselage to identify the flight.

31 EIA Christmas celebration for personnel and relatives inside one of the hangars at "Las Palmas" AB in December 1947.

Below:
Las Palmas, October 6, 1948. President Bustamante y Rivero reviews the crew and aircraft from 31 EIA at its home base after this unit participation during the so called "October revolt". (IEHAP)

EIA OPERATIONS DURING THE 1948 UPRISINGS

On July 4, 1948, the EP Cmdr. Alfonso Llosa Gonzales-Pavon, an officer who had become disaffected with the constitutional regime after government led Dr. José Luis Bustamante y Rivero had relaxed the pressure put over the APRA, natural enemy of Peruvian armed forces.[1] Considering this as an act of treason, Cmdr. Llosa led an uprising against the government. After learning about this situation the government ordered the CAP to take actions and, therefore, during the first hours of July 5 the *Comando de Operaciones* (Operations Command, COMOP), placed under direct command from the MA, ordered the deployment to "Alfredo Rodriguez Ballón" international airport, in Arequipa, of three

---

1   Armed forces, specially the Army, were notoriously anti-aprista since the slaughter of dozens of EP soldiers and officers by Aprista hordes during the 1932 uprisings.

Douglas 8A-5N seen undergoing a routine inspection to its engine at "Las Palmas" AB, in February 19, 1948.

Douglas 8A-5N serial 265 from 31 EIA being serviced by the unit ground personnel at "Las Palmas" AB on February 1948.

Douglas from 31 EIA aircraft, as well as a pair of Curtiss-Wright CW-22B Falcon from the 35 *Escuadrón de Información Terrestre* (35 EIT). Shortly after the trio of Douglas left "Las Palmas"AB and headed south bound to Arequipa, landing at that location nearly two and a half hours later. At Arequipa Cmdr. Luis Cayo Murillo, Chief in Command of the unit, decided to perform a reconnaissance flight over the city of Juliaca, after which he issued a complete report to the head of the army´s *IV Región Militar* (IV Military Region). With the available information and once last minute coordination had been held by radio with COMOP in Lima, Cmdr. Cayo received the order to attack the rebel forces and immediately began the planning of the following operations along the other two pilots, while ordering the ground personnel to ready their aircraft with bombs and ammunition.

At 1510hrs of July 5 the escadrille took off from "Alfredo Rodriguez Ballón" airport, with Capt. Jorge Camell del Solar, as flight leader, and Capt. Jorge Barbosa Falconí and 2nd Lt. Pedro Izquierdo Kernan as wingmen. An engine failure forced Capt. Camell and 2nd Lt. Izquierdo to abort the mission and head to "Alejandro Velasco Astete" airport in Cusco, designed as alternate airfield, while Capt. Barbosa continued with the mission, attacking the army barracks at Juliaca with bombs and machine gun fire. After completing their mission the pilot began its return to Arequipa but

An interesting picture of training aircraft lineup at "Las Palmas" AB during the year 1956. In the foreground, 8A-3P serial 280, sporting a nice cowling decoration, can be seen accompanied by its younger brother, an 8A-5N. Note the lack of machine guns on both aircraft.

A view from Douglas serial 276 in "Las Palmas" AB during taxiing before a routine flight. (IEHAP)

became momentarily disoriented due to the sun setting on his 12´, losing his direction. After flying for more than the calculated distance to Arequipa and with his fuel levels decreasing alarmingly, Barbosa began to worry and it was only the sighting of a geographic landmark, the Pomacanchi lagoon located at 10 minutes flight time from Cusco, the fact which allowed the pilot to correct his bearing and reach - with the last sunbeams- the airfield.

A few months later, the government faced a new crisis as the country fell in disarray during the so called "October revolution", which strongest manifestation was the so called Rebelión de la Armada, when, on October 3, 1948, rogue Peruvian Marina de Guerra (Peruvian Navy, MGP) elements took control of the main MGP bases and, more importantly, of the following warships: the cruisers BAP "Almirante Grau" and "Coronel Bolognesi"; the destroyers "Almirante Guisse" and "Almirante Villar"; the frigates "Teniente Ferré" and "Teniente Palacios" as well as other minor vessels. These forces, along thousands of APRA supporters, also took control of the "Real Felipe" fortress, in Callao, which armed themselves with the weapons took from the fortress armories. As in the past, Peruvian military aviation refused to take part in an uprising against the constitutional order, remaining loyal to the government. Consequently, the armed forces HQ ordered the CAP to, immediately, launch attacks against the dissident forces. The mission was trusted to the 31 EIA which at 0945hrs launched a seven aircraft formation, led by Cmdr. Enrique Escribens Correa, who took off from "Las Palmas" and headed west to the nearby Pacific Ocean searching for the rebellious fleet which was discovered stationed in front of the Miraflores, Barranco and Chorrillos districts. Escribens ordered his men to make mock up dive attacks over the ships trying to persuade their crews, but, during their first passage, the anti aircraft guns opened fire, fortunately without causing any losses. After returning to "Las Palmas" the unit awaited new orders, leaving again at 1130hrs with the mission to attack BAP "Admiral Grau", which at the time was refueling at the naval base located in El Callao harbor. After arriving over their objective, the flight began its bombing run as the ship aircraft fire opened fire on them, and as consequence all of the aircraft received impacts although none was critical. After completing their mission the aircraft returned to "Las Palmas" AB where close inspection by ground personnel showed multiple impacts on every single airframe, with Douglas serial *277* flown by 2nd Lt. Bohórquez carrying the worst part as it landed with both tires flattened due to bullet holes. A third sortie was launched at 1700 hours, flown by Capt. Barbosa, as leader, and 2nd Lieutenants Vásquez and Del Busto, as wingmen, with the mission to attack on rogue shipping sighted off San Lorenzo island, near Callao. After discovering their targets, the formation performed its attack launching their bomb load, four 50kg bombs each, in salvoes, achieving some near misses and damaging some of the ships.

Another view of Douglas 8A-5 N° 276 after it suffered a landing gear strut fracture during its landing at "Las Palmas" AB. The aircraft was repaired and flew again.

By the end of the day, the poorly supported "October Revolution" had been defeated, at the loss of thousands of civilians and military casualties, by night the MGP units returned to Callao and were surrendered by their crews.

Hardly two weeks after the bloody ending of the "October Revolution", the Peruvian government had to face a new uprising taken place near Arequipa, when, on October 27, army Division General Manuel Apolinario Odría radioed a manifesto informing the reasons behind his insurrection against the constitutional government. 31 EIA was once again called to action and two escadrilles left "Las Palmas" AB heading for Arequipa in order to serve as dissuasive force against the rebellious forces, but authorized to use lethal force if necessary. To the crew's surprise, while on their way to Arequipa the uprising had succeeded in removing the president Bustamante, and, as they descended from their planes at Arequipa airport, they were approached by armed guards who round them up and put them under custody. The CAP commander in situ ordered the removal of all bombs and ammunition from the six air-

craft as their pilots were jailed awaiting further orders. However, only a few hours later, the pilots were released and received orders to escort a CAP C-47 which carried to Lima General Odría and his staff. In a turn of destiny, the men trusted with defeating the uprising ended escorting the rebel leader now turned into president.

END OF THE LINE

After more than 15 years of service, wear and tear on their airframes, together with the lack of spare parts, began to take a toll on the 8A fleet and, in early 1958, the history of the Douglas model 8A in Perú finally came to an end when MA ordered the withdrawn from service of the last remaining 8A-3P and 8A-5 still in active duties with the Fuerza Aérea del Perú (Peruvian Air Force, FAP), by then serving as advanced trainers and photoreconnaissance aircraft. Two Douglas 8A have survived until our days: one preserved as gate guard at "Las Palmas" airbase, in Lima, while another, Douglas 8A-5, serial FAP *277*, rests over a pedestal placed on a main plaza of a small town near Andahuaylas, on the south eastern Andes region.

DOUGLAS 8A-3N
FOR THE DUTCH "MILITAIRE LUCHTVAART"

For many years, since the establishment of the Dutch Aviation Department, (Luchtvaartafdeeling) on 1 July 1913, military aviation was a small part of the Dutch armed forces. The armed forces as a whole, suffered under budget cuts after the end of the first World War. During the early 1930's very little funds were available for the Luchtvaartafdeeling.

In 1936, after many years of cutbacks, new funds finally became available for a large, four year spanning, modernisation plan for the Luchtvaartafdeeling. During 1938 plans were made for the purchase of a series of 18 two seater combat aircraft for the ground attack role in close cooperation with the Field Army. The following specifications were published:

- single engined high wing monoplane or biplane;
- climb to 3000 m (10.000 ft) within 5 minutes;
- top speed of 425 km/h (264 m/ph) at 3000 m
- maximum weight 3500 kg (7716 lb)
- armament: four machine guns firing forward, one trainable. Bomb load of four 50 kg (110 lb) bombs fitted to wing racks.

A tender was sent to domestic and foreign constructors, which led to eight responses, of which five were from Dutch constructors. Three offers came from foreign constructors, and were considered the most suitable. These were the Seversky 2PA, Douglas 8A and Caproni 134. There were some doubts about the Seversky firm, and the 2PA had yet to be tested. The Capro-

Main Dutch airfields
1   Soesterberg
2   Ypenburg
3   Ockenburg

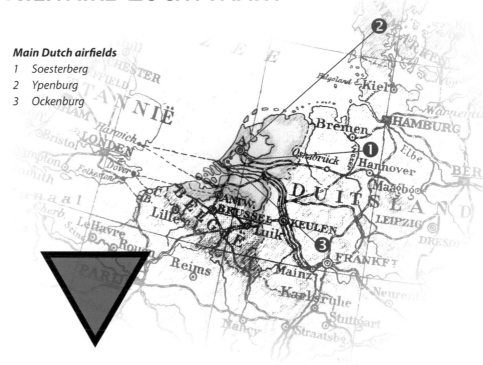

ni type had been available since January 1937, but development was terminated early 1939.

A Dutch committee, including a pilot and observer, visited the Douglas company, and a Peruvian machine was made available for a test flying program during 20 to 22 February. The Dutch navy attaché in the USA advised, based on the excellent flying qualities and technical background of the Douglas 8A, to order a batch of aircraft quickly. The Douglas could be delivered with a Wright Cyclone R-1820 G105A of 760 HP, R-1820 G103A of 872 HP or Pratt and Whitney S1CG of 1060 HP. Because the type was also ordered by other foreign parties, an offer was valid for a period of only a week.

The production line at El Segundo. Interestingly, the horizontal tail planes have received a coat of camouflage paint while the fuselages are painted with primer. (coll. F. Gerdessen)

Pilot capt. J.H. van Giessen, here at the left, was assigned to test fly a Peruvian machine during purchase negotiations. (Coll. E. Hoogschagen)

Observer capt. J.C. Kok, at right, was also involved in the test flying. (coll. F. Gerdessen)

The Dutch Minister of Defence replied on 3 March, confirming an order for 18 machines powered with the Pratt and Whitney engine at a price of $ 59350 per aircraft. The contracts were signed on 14 March. The Dutch machines were given type designation 8A-3N. The series was ordered including engines, spares, tools, blueprints and manuals, while 14 Fairchild gun cameras, bomb control panels, and five spare engines were ordered in the following period. Armament and radio equipment was not included. The Dutch would order 7,9 mm Browning machineguns at FN Liege, Belgium and fit these after arrival in Holland. Eighteen VR 67 radio sets were ordered from the Dutch company NSF. These would be built into the aircraft after arrival in Holland.

In order to ship the crated aircraft from San Pedro harbour on 26 August, construction and delivery of the machines

391 on the ramp at El Segundo. The aircraft were painted in a disruptive camouflage pattern of dark brown, green and sand. The dark brown was also applied to the underside of the wings and fuselage. (coll. F. Gerdessen)

was quickly started. The Dutch machines received construction numbers *531* up to *548* and military serials *381* up to *398*. The first machine, serialed *381*, was ready on 29 July and test flown by company pilot P.W. Coyle two days later. A test program followed from 11 to 24 August, after which the plane was accepted. Aircraft *382* and *383* were test flown on 22 and 23 August respectively. The first three aircraft were shipped on the 26th, just on schedule. The following batch of six aircraft, *384* up to *389* followed in September. Unfortunately, *386* ground looped during its test program and had to be returned to the factory for repairs. The *390* up to *398* were shipped between 8 and 14 November, and the repaired *386* on 16 November.

The first machines reached Holland on 21 October, and were disembarked in Rotterdam harbour. The rest of the machines followed during the following months.

The planes were planned to be part of 2 LvR (2e Luchtvaartregiment, 2nd Aviation Regiment), which had just been created. 2 LvR's operational task was to support the Field Army and was equipped with four reconnaissance groups (each group was given Roman digits I to IV). The reconnaissance planes should be protected by fighters, of the 5th group (V-2 LvR Jachtgroep Veldleger, 5th Fighter Group, 2nd Aviation Regiment). Four fighter departments were attached to V-2 LvR, two with single seaters, and two with two seater fighter aircraft.

The Douglas aircraft were intended for 3-V-2 LvR (3e Afdeeling, 3rd squadron). To bolster V-2 LvR's strength, purchase of a second batch of 18 aircraft was suggested, later even 120 machines. But this never materialized, because the opportunity arose to purchase a batch of 26 twin engine Fokker G.I Wasp aircraft. These were intended for Republican Spain, but embargoed by the Dutch government.

V-2 LvR was formed on 29th of August 1939, the day of the full Dutch military mobilization. First units were 1-V-2 LvR, with D.XXI fighters, and 2-V-2 LvR, with a stop gap equipment of obsolete Fokker D.XVII biplane fighters. 3-V-2 LvR followed, with reserve captain J.A. Bach as temporary commander. The Douglas aircraft were transported to Soesterberg airfield, where the workshops of the Luchtvaartbedrijf (LVB), the technical department, were located. During the assembly process, however, the LVB had to be diverted to a new location as part of mobilization.

Assembly of the first eight Douglas aircraft would be completed at the emergency airfield of Ockenburg

The airworthy aircraft where flown there, the rest followed by road transport. After a short period at Ockenburg, the aircraft returned to Soesterberg during the second week of December. Here, the final ten aircraft were assembled. During this period aircraft *383* received some damage to the starboard wing. The fuselage of *388* was damaged in January. In both cases, cause of the damage is not known.

After the aircraft had been erected and test flown, armament was built in during January and radio sets followed during the first week of March 1940. One machine was experimentally fitted with a direction finder, another with an alternative radio antenna, which gave better results compared to the original. The aircraft were also fitted with a propeller spinner.

Personnel was transferred from other units or directly from the flight schools. Some pilots had to receive additional training, while the crews were familiarized with the 8A-3N at Soesterberg airfield. An intensive training period followed. Douglas aircraft participated in mock ground attack exercises over the main defense lines on 23rd of February and during March similar exercises were flown, with AAA batteries as target. During one of these flights, *386* crashed at Ypenburg airfield on 15th of March. The crew, pilot sgt. vl. G. Nijhuis and observer elt.wnr H.F.J.M. Plasmans were killed. After examination, it was concluded that high speed stall was the most probable cause for the accident.

Following Germany's assault on Denmark and Norway on 9th of April, V-2 LvR was put on alert. Two patrols, of three and four Douglas aircraft, were formed and were on standby during the following days.

Soesterberg airfield was the cradle of Dutch military aviation. Here all 18 aircraft are lined up, although not all may have been made airworthy.
(coll. F. Gerdessen)

Auxilery airfield Ockenburg was quickly created by connecting a number of soccer fields.
(coll. F. Gerdessen)

Strength of 3-V-2 LvR counted nine aircraft in first line and three in reserve. The rest of the aircraft was in storage at the LVB. The units which were to be equipped with G.I Wasp aircraft never reached operational status. Most were still at the Fokker works when war erupted on 10 May 1940.

3-V-2 LvR was ordered to relocate to Ypenburg airfield on 7th of May. All 17 aircraft, two Fokker G.I's and three liaison aircraft flew to the new site the same day. Five of the Douglas aircraft were relocated to Ockenburg the following day and put in storage. Only a minimum of bombs, 72 bombs of 50 KG, was available at Ypenburg. The rest was sent to a depot, while a quantity

The Dutch roundels were replaced with large orange triangles following the shooting down of a Dutch plane on 13 September 1939. According to the German apology the Dutch markings were mistaken for the RAF roundel.

(Coll. E. Hoogschagen)

3-V-2 LvR remained at Soesterberg up to 7 May 1940. It then transferred to Ypenburg, which was located closely to the seat of government of The Hague. (coll. F. Gerdessen)

386 was lost during exercises. Low level flying was extremely dangerous and the heavy controls of the machines left very little room for error.

(coll. F. Gerdessen)

Neutrality markings included the orange triangles on fuselage and wings. The rudder was also painted orange. All markings were outlined with a 10 cm thick black line. (coll. F. Gerdessen)

of parachute bombs was still on order. 1-V-2 LvR, equipped with the Fokker D.XXI fighter, was also located at Ypenburg. II-2 LvR, a reconnaissance unit with a mixed equipment of Fokker C.V and Koolhoven F.K. 51 planes was dispersed around the edges of the field. From the 7th, all aircraft were on take-off alert from 03:15 to 08:00. After 08:00, half of the machines remained on alert.

WAR!

In the evening of 9th of May intelligence reported that the situation had become critical. Captain Bach was notified at 01:15 and immediately rushed to the field. All twelve airworthy aircraft were readied for combat and were standing by at 04:00. Bach separated the machines into four combat formations and gave each a radio call sign, Dutch boys' name Jan, Piet, Klaas and Lodewijk.

Six aircraft were not ready for combat;
Unserviceable: *383*
In storage: *394, 395, 396, 397 and 398*

A prewar photo of Ypenburg.
(coll. F. Gerdessen)

Technical data Douglas 8A-3N		
Engine	Pratt & Whitney GR-1830-SC3G	
Power	1065 hp	
Wingspan	47 ft 9 in (14,55 m)	
Length	31 ft 8 in (9,67 m)	
Height	9 ft 9 in (2,97 m)	
Empty weight	4,850 lb (2.200 kg)	
	Attack role	Light bomber role
All-up weight	3.580 kg	8,984 lb (4.075 kg)
Max. speed	254 m/ph (409 km/h) at 12,000 ft (3.660 m)	231 m/ph (373 km/h) at 12,000 ft (3.660 m)
Range		
Service ceiling	29,593 ft (9.200 m)	24,901 ft (7.590 m)
Climb rate	-	
Armament	4 fixed 7,9 mm mg in wings, one flexible 7,9 mm mg in rear cockpit	
Bombload	20 internally stored 17,6 lb (8 kg) bombs plus four externally carried 110 lb (50 kg) or 220 lb (100 kg) bombs or two 450 lb (200 kg) bombs or eight 110 lb (50 kg) bombs externally	

During the break of dawn, the noise of aircraft engines was heard, and at about 04:00 three unknown aircraft were approaching Ypenburg. The alarm was sounded and the order to scramble was given. The D.XXI's took off first, followed by the Douglas aircraft. While the last aircraft were taking off, the first bombs tore up the field. Each of the crews climbed to high altitude to gain an advantage, and after asking for instructions, all were ordered to attack. The patrols were almost immediately scattered, and aircraft got involved in dogfights individually.

PATROL JAN:

After engaging a Junkers Ju 52 transport, which was probably shot down, *382* was shot down. Pilot 2nd lt. Guijt successfully bailed out, lt. Vonk was killed.

Pilot sgt. D. Lub had to avoid bomb craters during take off with his machine, *384*. His plane was possibly damaged during the bombing, but was able to fly into combat. Above the Hague, the plane was fired at by friendly AAA. A Heinkel He 111 was engaged and gunner sgt. F. Vijn managed to score some hits, but when

The Dutch series were equipped with two-speed Hamilton Standard propellers. The Dutch national colours red-white-blue were applied on the tips.

(coll. F. Gerdessen)

Armament

The Dutch machines were armed with four fixed 7,9 mm Browning machine guns. Each wing housed two guns, with 500 rounds per gun. A simple visor was fitted in front of the wind screen. A fifth trainable 7,9 mm Browning gun was operated by the observer, who also acted as bomb aimer. The 8A-3N was fitted with a ventral gondola, which could be lowered manually. The bomb sight could be fitted in this gondola. The DB-8A-3N could carry a mix of various bombs. Eight bomb racks were fitted under the central wing panel for a total bomb load of 400 kg (two 200 kg bombs, two 100 kg bombs and/or four 50 kg bombs). Twenty 8 kg shrapnel bombs could be stowed vertically in an internal bomb bay. When these were carried, the maximum bomb load would be 560 kg.

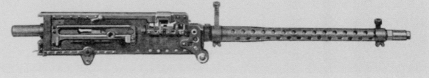

The Dutch machines were painted in a three-tone camouflage pattern of dark brown, green and sand. The bottom of the wings and fuselage was dark brown as well. (Profile by Alexandre Guedes)

Messerschmitt Me 110's appeared, Lub had no other option than to avoid an interception. By chance, the crew had a rendez vous with *381*, and the pilots agreed to land at the beach of Rozenburg. After a fight of nearly 1 hr. 15 mins. Lub force landed his plane, which roughly

Lt. Jansen managed to take off with *387* despite a damaged port wing. After climbing to higher altitude several German aircraft were attacked, with only the starboard guns actually firing. Suddenly the port wing collapsed, Jansen was flung out of the cockpit during the spiraling decent.

After a fierce struggle with the cockpit hood, gunner Beuving was able to parachute to safety as well.

PATROL PIET

The *388*, piloted by lt. van Riemsdijk, had to restrain from combat while the observer, sgt. Hagen, fixed a problem with his gun. When he was ready, *388* joined combat, but soon afterwards, Hagen was seen slumped over his gun. The machine was shot on fire by Me 110's, and van Riemsdijk had to hit the silk. He ended up hanging in a tree, his plane crashing nearby.

Sgt. Hinrichs was unable to enter the fight with *389*. After take off, the guns of his plane malfunctioned. After half an hour he landed at Ockenburg, where the plane was destroyed during ground combat.

Pilot sgt. de Bruijn and observer 1st lt. Van Boekhout managed to score hits on several enemy planes, and possibly shot down a Ju 52. *390* was damaged when intercepted by Me 110's and after shaking these with a steep dive, the plane was hit by friendly AAA. De Bruijn had to land with engine trouble in a field with cows.

Patrol Klaas

After expanding all ammunition, *391*, crewed by pilot 2nd lt. Heijen and observer sgt. Nijveldt, landed at Ockenburg. The plane burnt to the ground during the following fighting.

The *392* got involved in several fights, during which observer sgt. Staal shot a Ju 52 transport on fire. The plane was hit badly though, and both men used their parachutes safely.

During the fighting *393* was seriously damaged, and the crew, pilot 2nd lt. H. Pauw and cpl. L.M.J. Ballangée had to jump. Pauw was mortally wounded and died shortly after landing. The parachute of Ballangée failed to deploy, resulting in his death.

383 was unserviceable and was left at Ypenburg, where it was examined by curious crowds.
(coll. F. Gerdessen)

396, as it was found by the Germans at Ockenburg. A burnt out machine is seen in the rear.
(coll. J. van den Heuvel)

Patrol Lodewijk

After the fighting, 1st lt. Bierema attempted to land his plane, *381*, at the beach of Rozenburg, but landing had to be aborted. Bierema then flew to Kijkduin beach to try again. Before landing, the plane was intercepted and shot down by Me 110's. Bierema and his observer, 1st lt. Faber, were killed.

2nd lt. Scheepens and 2nd lt. Vermeulen were killed when their plane, *385*, was intercepted over Kijkduin.

During the fierce aerial combat, Ypenburg was heavily attacked and after a while, parachutists were dropped with the goal to conquer the field, so it could be used to fly in more troops. Ypenburg, just like nearby Ockenburg and Valkenburg airports where the ideal staging points to take the political capital of the Hague and capture the Dutch royal family. The Germans underestimated the defence of the field, which quickly turned into a chaotic scene of mangled airplane wrecks. The field was retaken by Dutch troops later on the day.

3-V-2 LvR was knocked out in an hour. Seven aircraft had been shot down, two landed at Ockenburg airfield and two elsewhere. Eight crewmembers had been killed, while three members of the ground crew perished. Six aircraft were in salvageable state, but one, registered *390* was plucked by souvenir hunters before the plane was transported away.

A Douglas, together with a Dutch Fokker D.XXI, was displayed at the Deutsches Luftfart Sammlung in Berlin. It was lost during the bombings of Berlin.
(coll. F. Gerdessen)

Right above and center:
An 8A, now operated by the Luftwaffe, at Brandenburg-Briest *(via F. Gerdessen)*

Below:
The Dutch colours were largely retained during the short use by the German Luftwaffe. The Dutch nationality markings were simply painted over and Luftwaffe markings were applied, as was usually done with captured foreign aircraft. *(Profile by Alexandre Guedes)*

During june 1940, at least two machines, former *394* and *395* were test flown at Erprobungsstelle Rechlin.
Two German registries are known, of planes flown with a Fluglehrerschule at Brandenburg-Briest. Of these, *DF+IR* crashed in August 1940, killing the crew. Another sample was displayed at the aircraft collection in Berlin, which was destroyed during the course of war.

The former 394 carrying the markings of Fluglehrerschule of Brandenburg-Briest (coll. J. van den Heuvel)

IRAQI AIRFORCE

Iraq ordered 15 aircraft, with type designation 8A-4. The machines were equipped with Wright GR-1820-G103 radial engines of 1000 HP and left the factory without armament. The aircraft were transported by ship to the Iraqi port of Basrah, where they arrived during September 1940. Armament for the planes was transported in a second vessel, but this was seized before its load could be disembarked.

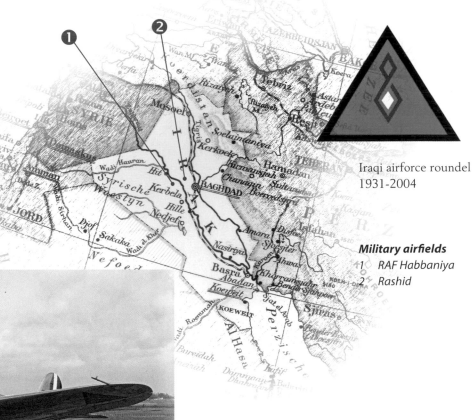

Iraqi airforce roundel
1931-2004

Military airfields
1   RAF Habbaniya
2   Rashid

Factory photos showing an Iraqi machine prior to delivery. Besides the nationality markings and a very striking emblem on the nose, the plane lacks an individual plane number. It seems that armament and radio still need to be installed.

(coll. E. Hoogschagen)

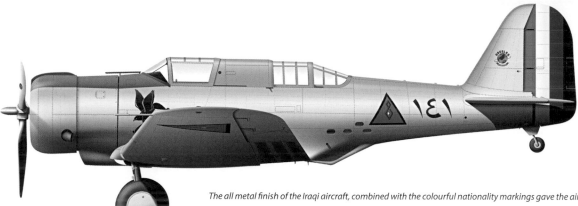

The all metal finish of the Iraqi aircraft, combined with the colourful nationality markings gave the aircraft a striking finish. It is unlikely that the aircraft received a coat of camouflage paint. The aircraft featured a stylized eagle carrying a bomb with inscription. The full text on the bomb is unfortunately not known.

(Profile by Alexandre Guedes)

On 3 April 1941 former prime minister Rashid Ali led a coup and installed himself as new prime minister. The newly formed government also pleaded for military assistance from Germany in case of war.

On 30 April, Iraqi forces, roughly 9000 men supported by artillery and armoured vehicles, took positions on high ground surrounding RAF Habbaniya, creating an encirclement of the camp. Iraqi forces also had the following aircraft available: (see table right)

RAF aircraft were still able to take off despite the siege. RAF Habbaniya was home to No. 4 Service Flying Training School and available aircraft were mostly trainer aircraft; 26 Airspeed Oxfords, 30 Hawker

Audax's and eight Fairey Gordons, complemented by nine Gloster Gladiators and a single Bristol Blenheim.[1] During the early morning of 2 May 35 aircraft of various types took off to attack dug in Iraqi forces around RAF Habbaniya. Wellingtons from

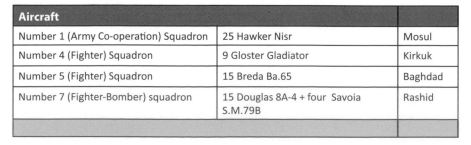

---

1   Sources are conflicting and mention different numbers of aircraft at hand.

Aircraft		
Number 1 (Army Co-operation) Squadron	25 Hawker Nisr	Mosul
Number 4 (Fighter) Squadron	9 Gloster Gladiator	Kirkuk
Number 5 (Fighter) Squadron	15 Breda Ba.65	Baghdad
Number 7 (Fighter-Bomber) squadron	15 Douglas 8A-4 + four Savoia S.M.79B	Rashid

37 and 70 squadrons joined after arrival at Basra. After aircraft returned, they were quickly turned around; refueled, rearmed, and repaired when necessary. After a debrief of the crew, a second crew received instructions and took over the plane for a new mission. The big Wellington bombers attracted the attention of AAA, but also of Iraqi aircraft. It was reported that two Iraqi Gloster Gladiator and two Douglas 8A's intercepted the bomber force, damaging a Wellington resulting in a forced landing. The campaign took over 7,5 hours of constant aerial action. At 10.00 am the Iraqi's launched a counter attack through air, destroying three RAF aircraft.

The main Iraqi Air Force aerodromes at Rashid and Baghdad were heavily attacked on 3 and 4 May, and again on 5 May, destroying workshops, hangars, fuel and ammunition depots and an estimated 22 aircraft, out of a force of roughly 50 machines. On May 6th the siege of Habbaniya was lifted and Iraqi forces were seen retreating. In the following hours all available aircraft were ordered to attack the Iraqi columns. Still, the Iraqi managed to launch an attack on Habbaniya. An aircraft, reported as being a Douglas 8A, was intercepted by a Gladiator and driven off.

After the siege of Habbaniya, the Iraqi Air Force had practically seized to exist. Soon afterwards, German and Italian air units entered the theatre.
It was however not the last time an Iraqi Douglas 8A took to the air. A single photo seems to suggest a surviving aircraft was still airworthy in October 1943.

Iraqi test pilot major Mahmoud al Hindi is portrayed besides one of the fifteen ordered aircraft.

(coll. E. Hoogschagen)

A TRAINER RÔLE IN LITTLE NORWAY

A factory photograph showing an assembled machine, still without armament. The machine carries temporary registry D-1. A lightning is painted on the length of the fuselage. These were red, white or blue.

When the Second World War broke out, the Norwegian air defences were in a neglected state. The venerable Fokker C.V biplane was still the most numerous aircraft at hand. This type, in use with the army air corps was built under license in Norway. The Naval Air Arm was flying a similarly outdated biplane type, the locally designed and built Høver M.F.11. After studying various options, the Vultee V-11GB was considered the best type for a modernisation and expansion programme. Sixty aircraft were needed, split over both services. A purchasing committee was sent to the United States but after arrival on 28 January 1940, it was found that the preferred type was not available for export. Committee members visited the Severs-

Left:
An unmarked machine during trials. Just like the Iraqi variant, the Norwegian machines were equipped with a Curtiss Electric variable pitch propeller. The aircraft remained unpainted, except for a dull black anti-glare panel in front of the cockpit. (via N. Mathisrud)

Right:
Aircraft 311, as it appeared before 1942.
(via N. Mathisrud)

ky, Curtiss, Vought, North American and Northrop factories (a subsidairy of Douglas). This resulted in the purchase of 36 Curtiss P-36 fighters and 36 Douglas 8A-5 aircraft.

The Northrop deal was signed at a total cost of $ 1.778.314,00, delivery would take place between September 1940 and January 1941. The aircraft were to be powered by Wright Cyclone GR-1820-G205A engines, giving a maximum power output of 1200 hp. Armament would comprise four fixed 7,92 mm FN Browning machineguns plus two additional 13,2 machineguns in underwing pods.

Before deliveries could start, Norway was invaded on 9 April. The German occupation was completed after Norway had surrendered on 10 June. A new destination for the aircraft was quickly arranged. A Norwegian flight school was to be found at Island Airport near Toronto, Canada. The first three machines were flown in on 2 November. These first aircraft carried temporary registries, *D-1*, *D-2* and *D-3*. The next two planes, *D-4* and *D-5* were ferried to Island Airport and left El

Norwegian training sites in Canada
1 Island Airport, Toronto
2 Muskoka Airfield, Toronto

319 and other DB-8A's on the line with P-36's. Just like earlier Norwegian aircraft, the series only received odd registries, painted in large black numbers on the fuselage tail. From late 1941 onwards, small registries were stencilled on the engine cowlings too.
(Forsvarets Museet via N. Mathisrud)

Segundo on 17 November. Unfortunately, *D-4* suffered an accident during landing at Albuquerque airfield and had to be returned to the factory. *D-5* arrived safely on 19 November. Another setback occurred on 14 December, when *D-6* belly landed at Mains Field. Delivery of this machine was also delayed. *D-7* to *D-10* arrived during December, while the repaired *D-4* and *D-6* followed in February 1941. The Curtiss fighters were also redirected to Island Airport.

All other aircraft were transported to Canada in crates, and remained crated. Condition of the aircraft was checked on 31 March and 1 April. During the same period, however, it was decided to train the Norwegian pilots on types they would encounter once in RAF service. This decision made the Douglas surplus, and on 22 May

A machine with gunpods and bombracks fitted. The bomb load could consist of eight 50 kg bombs or four 225 kg bombs.
(Forsvarets Museet via N. Mathisrud)

During September 1941 a group of trainees met with crown prince Olav, and his children prince Harald, seen here in the middle, flanked by his older sisters.

(Forsvarets Museet via N. Mathisrud)

the possibility of selling some of these aircraft was reported. Possible buyers were Peru or the Netherlands Indies. The sale of 18 machines to Peru was approved 4 days later, but when Peru got involved in a border conflict with neighbouring Ecuador, the export was embargoed by the United States. Instead, the aircraft were pressed into US service on 9 December.

The aircraft which remained were, together with the Curtiss fighters, used for advanced training. The training programme contained 27,5 flight hours on the Douglas, of which 7,5 were night flying hours.

311 at Island Airport. At left the tail of a Curtiss Hawk. Although the machines never reached Norway, they retained the Norwegian nationality markings, just like the other Norwegian aircraft. Because of this, the airfield gained the nickname "Little Norway". *(Forsvarets Museet via N. Mathisrud)*

Aircraft receiving attention on a summer day. In the middle 303, which was damaged beyond repair on 20 May 1943.
(Forsvarets Museet via N. Mathisrud)

The first course started in November and was concluded on 10 March 1941.

Up to February 1943, seven flying courses were completed. During the second course, five aircraft made a goodwill flight through the USA. The formation left Island Airport on 25 September 1941 and made stops at Bellafonte, Pensylvania, Bolling Field (Washington), La Guardia and Mitchell Field (New York). During their stay at La Guardia the crews had an audience with Norwegian crown prince Olav. All aircraft returned to Island Airport on 28 September.

116 students graduated during the courses. Four students lost their lives during crashes with a Douglas. Three machines were written off in accidents, while two more were scavenged for spare parts.

The training activities at Island Airport were completed in February 1943. The remaining thirteen airworthy machines, plus spares taken from *301* and *303* were sold to the United States, together with 12 Curtiss Hawk fighters. The planes were then sold to Peru under Lend Lease terms.

Aircraft		
313	5 July 1941	
327	20 January 1942	
301	14 September 1942	Crashed after engine fire. Disassembled and eventually sold as spares
321	24 January 1943	Crashed after entering spin. Pilot Sergeant O.H.M. Backer was killed
303	20 May 1943	Damaged in emergency landing. Disassembled and sold as spares.

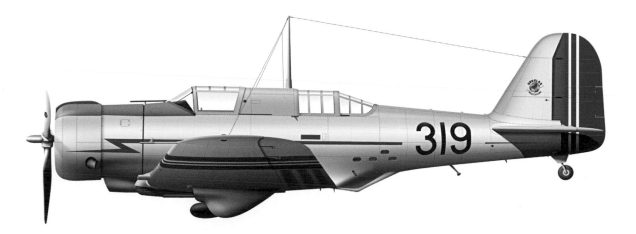

All Norwegian trainer aircraft which flew in Canada retained the Norwegian markings, consisting of flags in the national colours on wings and tail planes. (Profile by Alexandre Guedes)

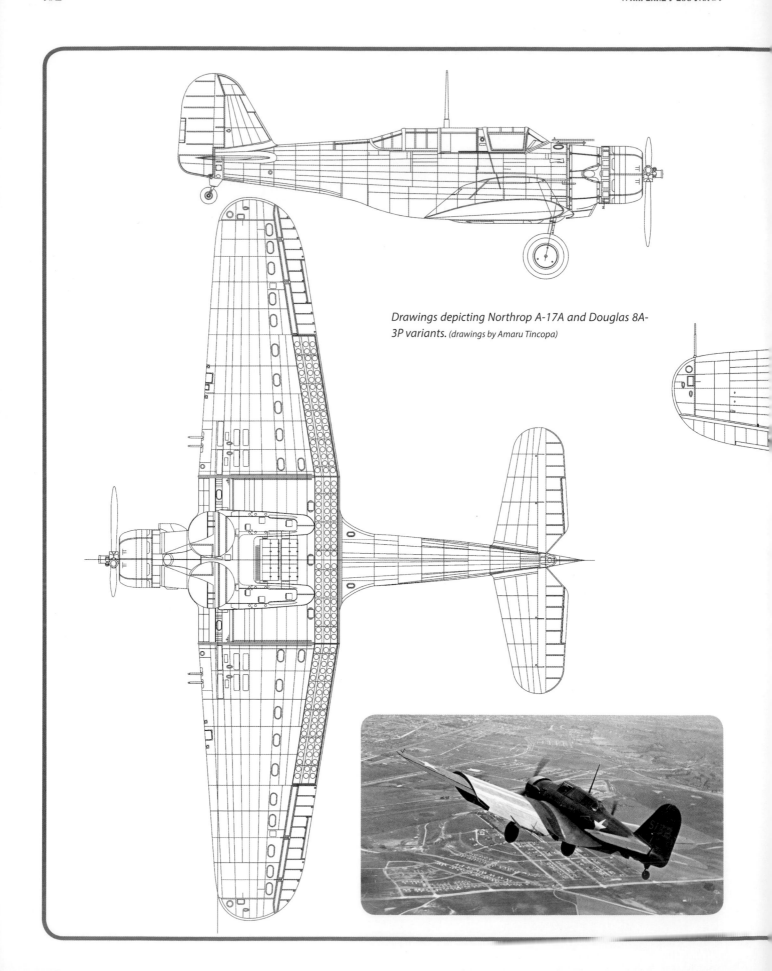

Drawings depicting Northrop A-17A and Douglas 8A-3P variants. (drawings by Amaru Tincopa)

Right:

The export aircraft carried company logos in various styles (drawings by Alexandre Guedes)

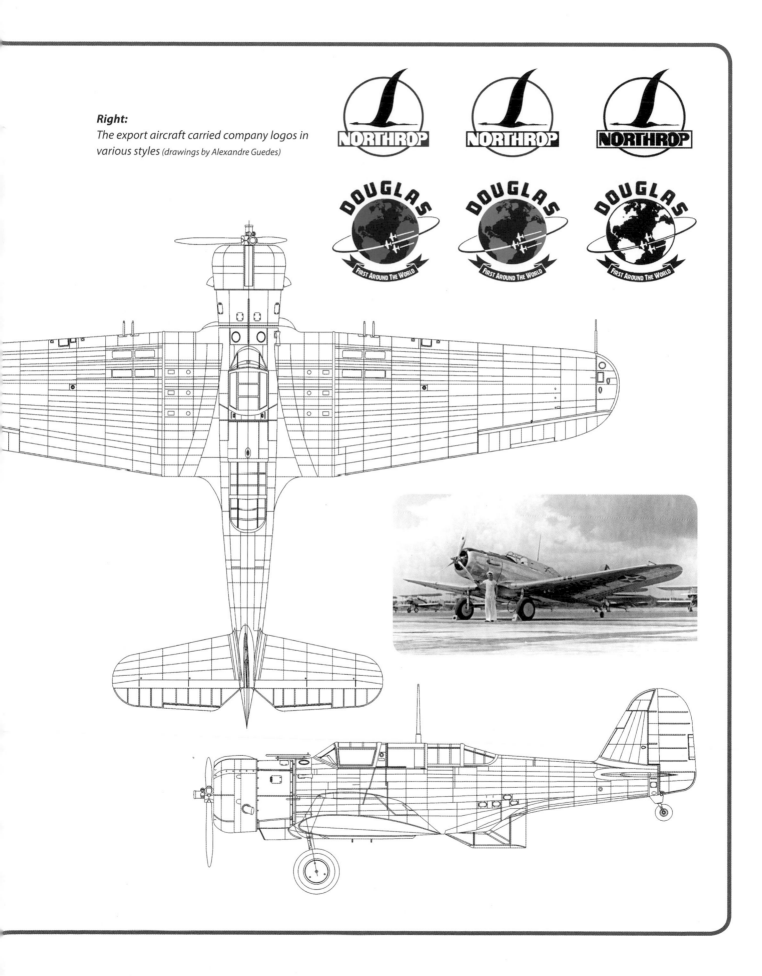

Sources

35-121 as it appeared between 6 May 1942 and 28 July 1943. It was written off in an accident on 17 May 1944. (Profile by Alexandre Guedes)

Books, articles, newspapers

- Bell, D.: *Air Force colors Volume 1 1926-1942*, Squadron/Signal Publications, 1995
- Bruin, de, R. et al, *Illlusies en incidenten*, Studiegroep Luchtoorlog, 1988
- Butler, P.: *Air Arsenal North America*, Midland Publishing, 2004
- Francillon, R.J.: *McDonnell Douglas Aircraft from 1920*, Volume I, Naval Institue Press, 1988.
- Gerdessen, F.: *Douglas DB-8A/3N*, Dutch Profile, 2012
- Griffin, J.A.: *Aircraft of the British Commonwealth Air Training Plan.*
- Hafsten, B. *Douglas 8A-5*, Norsk Luftfartshistorisk Magasin nr 3 2007 pages 12/18
- Hall. Å, Widfeldt, B.: *B5 Stortbombepoken*, Air Historic Research, 2002
- Handbook of service instructions – R-1535-7 -11 and -13 aircraft engines, May 20, 1936 (Revised 3-1-43)
- Heinemann, E. and Rausa, R.: *Combat aircraft designer*, Jane's Publishing (1980)
- Pelletier, A.J.: *Northrop's connection – part one*, Air Enthusiast May/June 1988 pages 62/67
- Pelletier, A.J.: *Northrop's connection – part two*, Air Enthusiast Sept/Oct 1988 pages 2/13
- Sanders Allen, R.: *The Northrop Story 1929 – 1939*, Orion Books, 1990
- *RAF batters Iraqi Air Force*, The Courier-Mail, Tuesday May 6 1941
- *Iraqi planes destroyed*, The Argus, Tuesday May 6 1941

Archives, websites

- Argentine Air Force and Argentine Army Annual memoirs, Public, Reserved and Secret Bulletins.
- Various. Historia Aeronáutica del Perú, Volumen IV, Instituto de Estudios Históricos Aeroespaciales del Perú. Lima, 1988.
- Various. Historia Aeronáutica del Perú, Volumen V, Instituto de Estudios Históricos Aeroespaciales del Perú. Lima, 1989
- Various. Historia Aeronáutica del Perú, Volumen VI, Instituto de Estudios Históricos Aeroespaciales del Perú. Lima, 1989
- Aviation Archaelogoical investigation & research database (www.aviationarchaeology.com)
- National Archives: AIR 54 - South Africa Air Force: Operations Record Books
- Air Corps Newsletter - issues spanning 1936 - 1941 (www.afhistory.af.mil/History/Air-Corps-Newsletter)
- Minutes of air council meetings 15, 23 and 27 at RCAF HQ, march-may 1941, via Directorate of history and heritage, Canadian armed forces
- Baugher, J.: USASC-USAAS-USAAC-USAAF-USAF Military Aircraft Serial Numbers - 1908 to Present (www.joebaugher.com/usaf_serials)
- Bell, K.: World War II: Air war over Iraq, Aviation History May 2004 (www.historynet.com/world-war-ii-air-war-over-iraq.htm)

Thank you:

First of all, I would like to thank my co-authors Nico Braas, Amaru Tincopa and Santiago Rivas for their support during realisation of this book. Without their knowledge about their fields of specialisation this book would not have been what it is now. Alexandre Guedes made the beautiful color profiles which added a very fine extra dimension to the book.

The following persons have contributed to the realisation of this book; Phil Butler, Ken Smy, Phil Scallan, Mathias Joost - thank you for your correspondence regarding British, Canadian and South African service. Rob Mulder, Nils Mathisrud, Mikael Forslund and Stig Jarlevik were very helpful during research on the Swedish and Norwegian histories. Frits Gerdessen provided many excellent photos, including some from the archives of late Jan van den Heuvel. John Shupek and Tony Chong assisted me while researching online sources on US Air Corps. Steve Donacik provided photo material from his private collection. I want to specifically thank mr. Gerald Baltzer, former Northrop employee and researcher, who kindly supported us with countless photos from his private collection.

Edwin Hoogschagen,
November 2019